OUR FUTURE EARTH

ALSO BY CURT STAGER

Field Notes from the Northern Forest

OUR FUTURE EARTH

THE NEXT 100,000 YEARS OF LIFE ON EARTH

CURT STAGER

DUCKWORTH OVERLOOK

First published in the UK in 2011 by
Duckworth Overlook
90-93 Cowcross Street, London EC1M 6BF
Tel: 020 7490 7300
Fax: 020 7490 0080
info@duckworth-publishers.co.uk
www.ducknet.co.uk

First published in the USA in 2011 by
Thomas Dunne Books, an imprint of
St Martin's Press, New York

A catalogue record for this book is available
from the British Library

ISBN 978-0-7156-4140-8

Printed in the UK by the
MPG Books Group

Contents

For Kary

Acknowledgments

The roots of this book run broad and deep, but the primary seeds that spawned it were the writings of journalist Elizabeth Kolbert and scientist David Archer. One of Kolbert's stories in *The New Yorker* ("The Darkening Sea") opened my eyes to the threat of ocean acidification from global carbon pollution, and Archer's research articles showed me that the time scales over which worldwide recovery from that pollution will play out are long enough to resonate with my own training in paleoecology. But those seeds also required fertile soil in which to develop, and many others have provided that.

My parents always encouraged my interests in the natural world, and many academic mentors helped them to mature into a profession. At Bowdoin and Duke, those included Janis Antonovics, Paul Baker, Dwight Billings, Larry Cahoon, Chuck Huntington, Art Hussey, Dan Livingstone, John Lundberg, Jim Moulton, Fred Nijhout, Steve Vogel, Henry Wilbur, and others.

While working on the story of the Lake Nyos gas disaster and other topics for *National Geographic* magazine, I learned much about science journalism under the guidance of Bill Allen, Tom Canby, Ford Cochran, Rick Gore, Chris Johns, and Tony Suau. Since then, I've been fortunate to continue that training with fine writer-editors such as Dick Beamish and Phil Brown (*Adirondack Explorer*), Betsy Folwell and Mary Thill (*Adirondack Life*), Maurice Kenny, and Christopher Shaw.

Shortly after joining the faculty at Paul Smith's College in 1987, I met another group of exceptional journalists at North Country Public Radio, which is based at Saint Lawrence University in Canton, New York. They invited me to provide a scientific complement to news director Martha Foley's role as interested lay person in a weekly five-minute conversation about everything from dragonflies to continental drift. First under the title "Field Notes" and then "Natural Selections," Martha and I have recorded hundreds of pieces over the last two decades, many of which are now archived online (www.ncpr.org). Thanks to Martha's patient but ruthless training, I now feel confident enough about explaining science to general audiences to enjoy it. Many thanks also go to Lamar Bliss, Ken Brown, Joel Hurd, Brian Mann, Ellen Rocco, and the rest of the NCPR team.

Many scientists contributed helpful ideas, feedback, information, and/or quotes to this project. Among them are Jun Abrajano, David Archer, Colin Beier, Dan Belknap, Paul Blanchon, Richard Brandt, Mark Brenner, Gordon Bromley, Ken Caldeira, Anny Cazenave, Brian Chase, Jeff Chiarenzelli, Brian Cumming, Ellen Currano, Kathie Dello, Andrew Derocher, Mike Farrell, Andrei Ganopolski, Gordon Hamilton, Darden Hood, Mimi Katz, Joe Kelley, George Jacobson, Andrei Kurbatov, Marie-France Loutre, Kirk Maasch, Paul Mayewski, Stacy McNulty, Mike Meadows, Johannes Oerlemans, Neil Opdyke, Kurt Rademaker, Don Rodbell, Bill Ruddiman, Dan Sandweiss, John Smol, Konrad Steffen, Gene Stoermer, Lowell Stott, Jerome Thaler, Piet Verburg, Chris Williams, Brendan Wiltse, and Kirsten Zickfeld. During the Copenhagen climate conference in December 2009, the American Geophysical Union organized a volunteer task force of scientists to offer a round-the-clock online resource base for journalists to approach with technical questions. I gratefully acknowledge the input of the following scientists who quickly and clearly answered many of my own arcane questions in that manner: Jill Baron, Jeffrey Dukes, Katharine Hayhoe, William Howard, Imtiaz Rangwala, Jeff Richey, Walter Robinson, Spencer Weart, and Bruce Wielicki. Any errors that remain in the text despite the efforts of these generous and capable experts are of my own doing, not theirs.

My research on African and Peruvian paleoclimates, some of which is summarized in this book, has been funded by the National Science Foundation, the National Geographic Society, and the Comer Foundation. In particular, Dave Verardo and Paul Filmer of the NSF have been great sources of encouragement for public outreach efforts as well as for research. Paul Smith's College, Saint Lawrence University, Queen's University, and the University of Maine's Climate Change Institute have also provided valuable support. My research on climate change in the Adirondacks and the Champlain Basin has been aided financially and logistically by the A. C. Walker Foundation, Paul Smith's College, and The Nature Conservancy.

Many friends, family, and associates have generously helped with editing, quotes, brainstorming, and/or other forms of support throughout this project, including Ken Aaron, Meg Bernstein, Sandy Brown, Pat Clelland, Lauralyn Dyer, Jorie Favreau, Kathleen Fitzgerald, Martha Foley, Eric Holmlund and his students, Kary Johnson, Devora Kamys, Hillarie Logan-Dechene, Bill McKibben, John Mills, Richard Nelson, Pat Pillis, Cheryl Ploof, Carl Putz, Mimi Rice, Christopher Shaw, Susan and Bill Sweeney, LeeAnn Sporn, Jay and Asha Stager, Mary Thill, and Will Tissot.

You probably wouldn't be reading this if Natalie Thill of the Adirondack Center for Writing had not invited scientist-author Bernd Heinrich to visit the Adirondacks on a speaking tour several years ago. After we first met and collaborated there, it was his kind endorsement that later landed this project in the capable hands of my agent-to-die-for, Sandy Dijkstra. Without Sandy's skill, energy, and top-notch staff behind it, along with the additional wizardry of agents John Pearce and Caspian Dennis, the manuscript might not have gained the attention of Thomas Dunne (Thomas Dunne Books, St. Martin's Press), Jim Gifford (HarperCollins Canada), Henry Rosenbloom (Scribe) and their counterparts at Duckworth/Overlook Press. It has been a pleasure to work with this team, particularly my excellent editor at Thomas Dunne Books, Peter Joseph. And both Meryl Moss Media Relations and Emma Morris (Scribe) worked wonders with publicity.

Above all, I am indebted to my best friend and partner-in-life, Kary Johnson, who has been a constant source of kindness and inspiration through these last three years of research, writing, and editing. Despite the piles of papers cluttering the house, the many evenings and weekends that saw me glued to a laptop, the last-minute requests for one of her beautiful photos or the drafting of a figure or help with a choice of words, and my general all-around obsession with this project, Kary has remained unceasingly helpful, insightful, and cheerfully tolerant of it all. Always willing to entertain a new idea or to help pursue a line of thought, always there when I needed someone to watch my back, she has made this work worth pursuing, and I could never have done it without her.

OUR FUTURE EARTH

Prologue

Out of the earth has come a creature that has
changed . . . the face of continents, that has
harnessed the forces of the earth and turned
*   them against themselves.*

        —John Burroughs, Accepting the Universe

Welcome to the Age of Humans, a new chapter of Earth's history whose name has already entered the lexicon of mainstream science.

Welcome to the end of the natural world as a realm that is somehow meaningfully distinct from humanity, thanks in large part to the worldwide carbon pollution that you and I have unwittingly helped to create and that will affect our descendants for many thousands of years, far longer than most of us yet realize.

And welcome to this peek beyond the curtain of 2100 AD, which currently marks the outer temporal limits of most thought and debate about modern climate change. As you'll soon see, the environmental consequences of our actions today are so large, powerful, and long-lived that they cannot be fully understood from a mere century-scale point of view.

My aim in these pages is to introduce you to a broader perspective on global warming than the one most readers are familiar with, the one that considers "long-term" climate change to be a trend that merely stretches over several years or decades. People like Bill McKibben and Al Gore have brought the planetary scope of CO_2 pollution to the attention of millions, but for most of us, the element of time has yet to be fully explored.

David Archer, a farsighted climate modeler whose work I will introduce later in the book, has described the situation thus: "The idea that anthropogenic CO_2 release may affect the climate . . . for hundreds of thousands of years has not yet reached general public awareness." The time has come to move on from unproductive, politicized arguments over global warming to the next stage of inquiry, where it is no longer a question of *if* it is happening but of when, how much, and for how long.

At first, it might seem strange that an environmental historian, or paleoecologist, like myself should be writing about future events like this. I read the stories of ecosystems that lie stacked in archives not of paper but of mud. My specialty is collecting layered core samples from the bottoms of lakes and bogs in the Adirondacks, Peru, and much of Africa, picking out the remains of microscopic organisms that once lived and died there and reconstructing past climates from the patterns of change they reveal. A layer of salt-loving algae tells me that the local climate was once dry enough to lower the lake level and turn the water brackish. A slug of pollen in another layer attests to wetter conditions that favored forests over deserts.

What a paleoecologist contributes most naturally to this prediction business is a sense of time. Much of what lies ahead of us has already happened before, and those of us with a long-term view of environmental history can often recognize familiar age-old processes in action today as well as likely consequences that may come as a result. We who combine the biological and geological sciences in our historical research also become used to thinking in broad terms that include both the living and nonliving worlds. But more to the point, we also think "deep." For us, a century or millennium may be just an appetizer on the menu, and the duration of a single human lifetime is, statistically speaking, insignificant.

That's not always a popular position, of course. A long view is not necessarily welcome to those who are preoccupied with events in the here and now, but it nonetheless offers potentially useful compass bearings for navigation in a complex and changing world. As a guide on this tour of the future, I'll look beyond the present moment to focus both forward and backward in time, bringing into view the nature of things to come as well as things that have long since been.

The relatively few scientists who have looked deeply into our future like this see the lingering climatic and ecological effects of fossil fuel carbon stretching well beyond the end of the twenty-first century. In order to follow their lines of sight through these pages, we will need to train the mind's eye to take in tremendous sweeps of Earth history, both past and future. Much of what we learn here will come from geoscientists who speak of eras and epochs as the rest of us speak of seasons, people who share with the superrich a close working familiarity with the significance of terms like "million" and "billion." To these professional time specialists, the Eocene and Pleistocene epochs are as real as World War II or the turbulent 1960s are to the rest of society, and they see in those long-gone ages some important lessons that can guide us as we struggle to understand what is happening around us today.

Before we go further in this quest, though, I would like to introduce you to a newly minted technical term that's currently striking resonant chords in the scientific community. At the moment, it still sounds foreign to most ears, but for those who are familiar with it the word represents an almost thrilling acknowledgment of our place in the grand arc of geologic time. It's modeled in the form of other major subdivisions of the fossil record that include what some call the Age of Fishes or the Age of Dinosaurs. Now that human influence has touched almost every cranny of the Earth, a new age has dawned, and it needs a formal name.

In the arcane lexicon of geological nomenclature, this age best qualifies as the latest in a string of episodes known as "epochs" that began 65 million years ago with the demise of the dinosaurs. You might have heard of them before, if you've spent much time learning about fossils. The postdinosaur years began with the warm Paleocene, whose prefix *paleo* (Greek for "ancient") refers to its great age relative to others that came later. Next came the even warmer Eocene, which saw the early-morning stages of modern mammal evolution. After skipping through three more epochs (Oligocene, Miocene, and Pliocene), each with a distinctive story of evolving life to tell, we find the Earth cooling down during the Pleistocene, and come, at last, to the climatically mild Holocene ("recent whole"), which began 11,700 years ago and traditionally includes the ever-advancing present.

But keep on the alert for a new epochal name that was recently bestowed upon this Age of Humans. No, it's not the "Plasticene," though blogger Matt Dowling has indeed proposed that label with tongue planted firmly in cheek. Partial credit for promoting the new name goes to atmospheric chemist and Nobel laureate Paul Crutzen, but it actually originated with aquatic ecologist Eugene Stoermer. Now an emeritus scholar at the University of Michigan, Stoermer recently told me that his catchy term spread informally through the scientific grapevine before appearing in print several years ago under the joint authorship of Crutzen and himself.

"I can't remember exactly how it first came to mind," he recalled, chuckling like a pleased but somewhat surprised parent whose kid has grown up to become a celebrity. "I used it in conferences here and there, and it eventually caught people's attention." That's not surprising, though, because Stoermer's term neatly defines these and near-future times as indelibly marked by anthropogenic, or human-generated impacts, and it's seeping more and more comfortably into the writings and speech of scientists and lay folk around the world.

So here's your chance to impress your friends, if you're the type who likes to show off by using the latest technical jargon that describes not only recent human history but also the next hundreds of thousands of years of it that are yet to come. Tell them, in the course of casual conversation, "Welcome to the Anthropocene." (Stoermer pronounces it ANthropocene, but anTHROPocene also works.)

By most definitions, the Anthropocene began during the 1700s when our greenhouse gas emissions started to change the atmosphere significantly. But our influences actually extend far beyond climate alone. The formerly dark portion of Earth that faces away from the sun now glows with electric light, as if it were illuminated by billions of fireflies. According to Crutzen, our fishing industries remove more than a third of the primary productivity of temperate coastal areas every year. Farmers sprinkle, spray, and spade more nitrogen fertilizer than is naturally deposited on all the world's forest floors, savanna turf, and bird rookeries combined. And species extinctions today are beginning to outpace any in the history of life.

A small but active corps of visionary scientists is now sketching the broad outlines of what the Anthropocene holds in store for us. But before looking further into the surprising details of what is to come, it's worth noting that we are not the only living things to have changed the atmosphere so much. From a biologist's emotionally detached perspective, there is nothing particularly unusual about the human tendency to pollute our environment; every organism produces waste, and the more organisms that exist in a given habitat the more unwanted by-products they produce. It's just that we humans have now become so numerous, so widespread, and so adept at consuming natural resources that our wastes are polluting the entire planet, even to the point of changing its climate. In that sense, we're becoming victims of our own success as a species.

The first such global pollution crisis was actually the work of marine bacteria, and it struck just over 2 billion years ago at a time when all life on Earth was single-celled. The pressures of mutation and natural selection drove some pioneering microbes to overuse a new way of harnessing the energy of sunlight—what we now call photosynthesis. Unfortunately for most of the other diminutive life-forms of the time, that primordial biotechnology also released a dangerous waste gas into the surroundings. That waste gas was free oxygen.

Excess oxygen steadily polluted the oceans as they grew greener and greener with the tint of chlorophyll and the atmosphere grew more and more corrosive as a result. Formerly gray or black rocks crumbled into reddened remnants of their former selves as the iron particles within them rusted. Any species that could not repair the ravages of oxidation in their cells perished or lived imprisoned in protective aquatic muds. Descendants of those microbial refugees still cower in the fetid muck of marshes and in the oxygen starved depths of certain lakes and seas. We unwittingly harbor legions of benign oxygen haters in the dark recesses of our digestive tracts, and some legumes such as soybeans pack their root nodules with blood-colored, oxygen-binding compounds that shield their resident bacteria, thereby earning paybacks in the form of microbial nitrogen fertilizer.

If language had existed back then in that purely microbial world, headlines would have heralded the advent of a global oxygen catastrophe. Perhaps bacterial alarmists who warned of that first pollution disaster would have described us humans as monstrous, two-legged versions of the "cockroaches that will take over the world after it's been poisoned."

In fact, both our distant ancestors and those of modern cockroaches did indeed populate the world only after photosynthetic oxygen made it habitable for animal life.

High above the early oceans, the novel molecules spawned new by-products just as the chemical stew of modern smog does today. Oxygen in the upper reaches of the atmosphere clumped into heavy, tripled clusters and accumulated as a layer of invisible ozone that blocked much of the sun's dangerous ultraviolet radiation. Meanwhile, down below, some of the primitive single-celled survivors of the oxygen pollution crisis were developing ways to use the poisonous gas as a source of energy in its own right. Eventually, Earth's first wriggly protozoans learned to harness oxygen's destructive power to convert the bodies of their smaller neighbors into useful food, and the rest is predatory history.

Today, the waste gas of photosynthesis contaminates a fifth of the air in our lungs and we, the descendants of those first polluters, can't live without it. When the world changes so dramatically, there must always be winners and losers. In this case, we have clearly been among the winners.

About a billion and a half years after the oxygen crisis, early plants that inherited the solar technology of photosynthesis were turning it to new uses of their own. Where the power of sunlight once supported only singular, free-living cells, increasingly large and abundant land plants used it to coax CO_2 molecules from the air, dissect them, and bind their carbon atom components into the fabrics of branches, trunks, leaves, seeds, and spores.

Growing atom by atom, like living crystals, primeval swamp forests hoarded precious carbon. Photosynthetic life plucked CO_2 from a thin gaseous soup in which the target element, carbon, was outnumbered more than ninety-nine to one by oxygen and nitrogen. At death, they took the concentrated carbon troves to their watery graves and were buried, layer upon layer, in mausoleum vaults of mud.

Hundreds of millions of years later, the first hints of Stoermer and Crutzen's Anthropocene began with another biogenic pollution event. Our industrial ancestors unearthed some of those black fossil deposits, called them coal, and set fire to them. Heated in the presence of oxygen, the purified carbons disintegrated back into diffuse swarms of CO_2 molecules, unleashing the hot solar energy of countless Paleozoic summers as their ancient chemical bonds snapped and cast them skyward.

Though at first indistinguishable from the other CO_2 molecules circulating among plants, animals, waters, and winds today, these fossil fumes are different. Most of the CO_2 that enters the air from breath, forest fires, oceanic upwellings, and rot is quickly recycled; about as much carbon is absorbed by photosynthetic bacteria, algae, and plants each year as is released by respiration, and roughly as much of it dissolves into the ocean surface as is naturally degassed from it. At the global level, only a small fraction is lost to sediment burial over the course of a year and only relatively modest amounts hiss from volcanic vents, so the total amount in circulation normally varies little.

Fossil fuel carbons, in contrast, are outsiders. Though some manage to rejoin the ebb and flow of modern life, most join the ranks of the footloose unemployed, swelling the pool of airborne CO_2 faster than other processes can reduce it. Just before the dawning of the Anthropocene, a random sample of a million air molecules would have netted you about 280 carbon dioxides. As I write this I could land 387 or so, many of which emerged from smokestacks and tailpipes within the last 250 years.

Why does this modern pollution spree deserve a new, formal geological name? Even though it represents less than 1 percent of the gases in the atmosphere, the growing surplus of CO_2 is now making the world hotter than it would otherwise be. Likewise, geologists designated the last two epochs largely on the basis of their climatic conditions; the Pleistocene was dominated by numerous glacial coolings and the Holocene was the latest of several shorter interglacial warm spells, the one during which the first complex human civilizations were born.

As I'll explain later, the greenhouse gas pollution of the Anthropocene will hang around long enough to cancel the next ice age, and the result is that this human-driven epoch may last an order of magnitude longer than the Holocene did. Incredibly, it is we—specifically those of us who live in the twenty-first century—who will do the most to determine its duration. The epochal name is well chosen; this Age of Humans is the product, the environmental backdrop, and the geological trademark of our species.

To some, the Anthropocene marks the end of nature as an entity separate from the apelike *Homo sapiens* species that it spawned in Africa long ago. Much of this conception of humans as privileged occupiers of some lofty plane above other species dates back to Aristotle's *Scala Naturae,* which is often translated as "The Great Chain of Being." It pictures a ladder or interlocking chain of existence that positions more complex animals above simpler ones and a heavenly creator above all. Because humans in this view combine both physical and metaphysical traits, they form a unique link that joins the celestial and earthly realms. Vestigial traces of the concept still linger in biological nomenclature that classifies complex-looking orchids as "higher plants" and simple-looking mosses as "lower plants." In society at large, it crops up in such terms as "the missing link," the theoretical hairy ape-human that would forge a lowly, anchoring ring in the great chain between us and other primates.

To most biologists today, however, the idea that humans are meaningfully separate from nature is rather old school. Our very ability to change climate on a global scale, simply by emitting our daily wastes, attests to our intimate connection with our physical surroundings. One could even argue that this kind of self-centered and shortsighted conceit, the idea that we are somehow exempt from the ancient laws of the physical world, is what got us into so much trouble in the first place.

This brings us back to an aspect of the Anthropocene revolution that is still under debate in the scientific community. When did the new epoch actually begin? Crutzen and others like him who focus on industrial emissions typically choose the mid- to late 1700s as that starting point. Some tie it specifically to James Watt's development of the modern steam engine in the 1760s.

Others, like climate historian Bill Ruddiman, put it thousands of years earlier. Ruddiman's idea helps to explain a mysterious anomaly in the record of ancient greenhouse gases that is preserved in air bubbles trapped in deep glacial ice. Ice cores, from Greenland and Antarctica, represent hundreds of thousands of years of climate history, and they reveal an intimate connection between past climates and greenhouse gases. These polar ice records show that while climates have seesawed violently between frosty ice ages and warm interglacials in the past, equally dramatic shifts in carbon dioxide and methane concentrations have also occurred, most of which, as we'll see in later chapters, had nothing to do with human activity.

Through most of that history, atmospheric concentrations of these two greenhouse gases fluctuated in near lockstep with each other, but something odd happened during the warm Holocene epoch, which began with an abrupt end to the last major cold episode 11,700 years ago. After an early thermal peak, temperatures began to slide back down into a long-term cooling trend. However, about 8,000 years ago, the CO_2 content of the air began to rise again instead of falling, as it had normally done during cool-offs of the distant past. Several millennia later, methane lifted off independently, too. Ruddiman proposes that the unusual CO_2 rise reflected widespread forest burning and land clearance for agriculture, and that methane later rose in response to the spread of Asian rice production in artificial, gas-bubbling wetlands. In that case, human impacts on world climate might have begun as early as 8,000 years ago.

Still others argue that climatic effects should not be the only criteria for tracking the history of human impacts on Earth. Most biohistorians believe that Stone Age hunters exterminated mastodonts and giant ground sloths along with many other large mammals roughly 10,000 to 15,000 years ago, and their disappearance fundamentally and artificially altered ecosystems all over the planet. In North America alone, more than half of all mammal species weighing more than 70 pounds (32 kg) vanished, and those weighing more than a ton (900 kg) were completely wiped out. One could therefore make a logical case for omitting the Holocene epoch from the geologic time scale altogether and simply folding it into the Anthropocene.

But most of us are less interested in when the Anthropocene began than in what it's going to be like from here on out. Just as fossils and ice cores give us glimpses into the world as it once was, the new science of long-term climate prediction sketches a compelling outline of things to come. In that expansive view, the basic shape of the future already exists, and we can use it to tell the full story of carbon pollution from start to finish rather than settling for the relatively short portion that now dominates our collective thinking. The pacing of most of these coming events will be sluggish on the scale of daily human experience, but their eventual cumulative effects on ecosystems and societies will be enormous and incredibly long-lasting.

And just what is it going to be like from here on out? We'll have to wait for time itself to reveal the details of future political systems, technologies, social interactions, and lifestyles; one never really knows what *Homo sapiens* will do next. But many features of the physical world are far more predictable. This book offers an introduction to those aspects of long-term climatic and environmental change that stand most clearly before us on the horizon. Here is a sampler of what is to come.

We face a simple choice in the coming century or so; either we'll switch to nonfossil fuels as soon as possible, or we'll burn through our remaining reserves and then be forced to switch later on. In either case, greenhouse gas concentrations will probably peak some time before 2400 AD and then level off as our emissions decrease, either through purposely reduced consumption or fossil fuel shortages. The passing of the CO_2 pollution peak will trigger a slow climate "whiplash" in which the global warming trend will top out and then flip to a long-term cooling recovery that eventually returns temperatures to those of the preindustrial eighteenth century. But that process will last for *tens or even hundreds of thousands of years*. The more fossil fuel that we end up burning, the higher the temperatures will rise and the longer the recovery will take.

There's much more to CO_2 pollution than climate change, though. Carbon dioxide will gradually acidify much or all of the oceans as they absorb tons of fossil fuel emissions from the air. That chemical disturbance threatens to weaken or even dissolve the shells of countless corals, mollusks, crustaceans, and many microorganisms, and their loss, in turn, will threaten other life-forms that interact with them. In some ways, this situation resembles the contamination of the primordial atmosphere by microbial marine oxygen, only in reverse; we are responding 2 billion years later with a corrosive gas of our own that is moving from the air back into the sea. Eventually, the neutralizing capacity of Earth's rocks and soils will return the oceans to normal chemical conditions, but the acid-driven loss of marine biodiversity will be among the most unpredictable, potentially destructive, and irreversible effects of Anthropocene carbon pollution.

Before the end of this century, the Arctic Ocean will lose its sea ice in summer, and the open-water polar fisheries that develop in its absence will last for thousands of years, radically changing the face of the far north as well as the dynamics of international trade. But when CO_2 concentrations eventually fall enough, the Arctic will freeze over again, destroying what will by then have become "normal" ice-free ecosystems, cultures, and economies.

Much or all of Greenland and Antarctica's ice sheets will melt away over the course of many centuries, with the final extent of shrinkage dependent upon how much greenhouse gas we emit in the near future. As the edges of today's icy coverings draw back from the coasts, newly exposed landscapes and waterways will open up for settlement, agriculture, fishery exploitation, and mining.

Sea level will continue to rise long after the CO_2 and temperature peaks pass. The change will be too slow for people to observe directly, but over time it will progressively inundate thickly settled coastal regions. Then a long, gradual global cooling recovery will begin to haul the waters back from the land. But that initial retreat will be incomplete, because so much land-based ice will have melted and drained into the oceans. At some time in the deep future, the sea surface will come to rest as much as 230 feet (70 m) above today's level, having been trapped at a new set point that reflects the intensity and duration of the melting. Only after many additional millennia of cooling and glacial reconstruction will the oceans reposition themselves close to where they lie now.

We have prevented the next ice age. The ebb and flow of natural climatic cycles suggests that we should be due for another glaciation in about 50,000 years. Or rather, we used to be. Thanks to the longevity of our greenhouse gas pollution, the next major freeze-up won't arrive until our lingering carbon vapors thin out enough, perhaps 130,000 years from now, and possibly much later. The sustained influence of our actions today on the immensely distant future adds an important new component to the ethics of carbon pollution. If we consider only the next few centuries in isolation, then human-driven climate change may be mostly negative. But what if we look ahead to the rest of the story? On the scales of environmental justice, how do several centuries of imminent and decidedly unwelcome change stack up against many future millennia that could be rescued from ice age devastation?

These are the sorts of extraordinary things that you'll encounter in this book, but rest assured that it's not just a litany of gloom. I hope instead to leave you with a well-founded sense of hope and a wake-up call. You and I are living in a pivotal moment of history, what some have called a "carbon crisis"—a crucial and decisive turning point in which our thoughts and actions are of unusually great importance for the long-term future of the world. But all is not yet lost, and climate change is not on the list of deadly dangers to most humans; as I will explain later, *Homo sapiens* will almost certainly be here to experience the environmental effects of the Anthropocene from start to finish. And that's only fitting, seeing as we're the ones who launched this new epoch in the first place.

But why, then, should we care enough about the distant future even to finish reading about it on these pages? The reason is simple. Although humans will survive as a species, we are faced today with the responsibility of determining the climatic future that our descendants will live in. It may well be a struggle to hold our carbon pollution to a minimum, but failing to take the heroic path and control our collective behavior is likely to drag us and our descendants into a realm of extreme warming, sea-level rise, and ocean acidification the likes of which haven't been seen on Earth for millions of years. And the outlook for most non-humans is far more worrisome than it is for our own kind. Severe environmental changes have happened before, even without our influence in the mix, but the situation that we and our fellow species now face is unique in the history of this ancient planet.

So welcome to this glimpse of our deep future. Welcome to the Anthropocene.

1

Stopping the Ice

One can only hope that the expected extremes of the Anthropocene will not lead to conditions that cross the threshold to glaciation.

—Frank Sirocko, paleoclimatologist.

Shockingly long-term climatic changes await us as a result of modern human activity, but examining our effects on the deep future also raises a related question that is well worth considering: what would global climates have been like if we had left our fossil fuels in the ground rather than burning them?

In that alternative reality our descendants would still fret about climate, sea levels, and ice caps but the news would read quite differently from that of today. "There's a massive, destructive climatic change coming, but scientists say that we can stop it if we take appropriate action now. If we go about business as usual, coastal settlements will be destroyed by sea-level shifts and entire nations will be covered with water. Frozen water. But there's still hope. If we simply burn enough fossil fuels, we'll warm the atmosphere enough to delay that icy disaster for thousands of years."

I'm talking about the next ice age. When a paleoecologist like myself thinks about global climate change the exercise is as likely to involve visions of ice-sheet invasions as it is to include greenhouse warming. We still don't know exactly why continent-sized glaciations come and go as they do, but they clearly have a rhythmic quality to them. Natural cyclic pulses take the long line of temperature history and snap it like a whip, looping it into a series of steep coolings and warmings. When viewed from a long-term perspective, major warmings of the past 2 to 3 million years can seem like brief thermal respites when the world came up for air between long icy dives; that's why we call them "interglacials" rather than something that sounds more normal or permanent. The cyclic pattern also suggests that more ice ages await us in the future, so strongly in fact that climate scientists routinely refer to our own postglacial warm phase that we live in today as "the present interglacial." Because of this admittedly unusual perspective, many of the paleoecologists I know balance their concerns about modern climate change with "yes, but it could also be a lot worse."

Although such views are rare outside of narrow academic circles, I believe that they belong in the mainstream. Time perspectives long enough to include ice age prevention are not just the stuff of mind games but potentially important aspects of rational planning for our climatic future. In order to appreciate why this is so, however, it helps to look more deeply than usual into the nature of ice ages.

The last one began about 117,000 years ago and ended 11,700 years ago. During that long and terrible reign of cold, roughly a fifth of the world's land surface resembled the icy interiors of Greenland and Antarctica today, especially in the higher northern latitudes. Most of what is now Canada and northern Europe was smothered under immense sheets of slowly creeping ice up to 2 miles (3 km) thick. The sites of today's Chicago, Boston, and New York were obliterated, and what we now call Long Island is a plowed-up bow wave of detritus that marks the southern limit of the last major ice advance. Entire landscapes sagged under that tremendous weight, pressing down hundreds or even thousands of feet into the planet's softer innards, and the gritty underbelly of the ice gouged deep scratches and grooves into solid bedrock that still scar the formerly glaciated regions of the world.

When you see glacial deposits and ice-scoured rock formations along a northerly roadside or trail, it's easy to let your imagination strip away the towns and trees and crush your surroundings under great, grinding slabs of ice. I envision it quite often near my home in the Adirondack Mountains of upstate New York. Recently I was reminded of the frozen past when I stepped off a woodland path near Saint Regis Mountain to take a closer look at one of the largest glacial erratic boulders I've ever seen.

The massive chunk of gray anorthosite was broader and taller than my house, and the prying fingers of winter frost had plucked garage-sized flakes away from its lichen-crusted flanks. They lay in low heaps around the central body of rock like cast-off clothing. The base of the giant perched just high enough above the ground to leave shadowed crawl spaces below that made me think of of crouching hermits and cave bears. Peering into one of them I scanned its dusky floor for signs of residents but saw only earth-colored gravel. Clean, well-sorted, smoothly rounded gravel, just like the stuff in the shallow streambed nearby. Gravel that was never buried under forest soils or leaf litter and that still looked as fresh as it did when melting ice dropped this gigantic sheltering rock on top of it.

That primeval scene drew my imagination back to when these mountains were still emerging from their long, lightless imprisonment. The rustling beech, maples, and birches before me faded away, along with the duff and dirt beneath them, exposing a desolate brown wasteland of

The author examining a glacial boulder near Saint Regis Mountain in the Adirondack Park, upstate New York. *Kary Johnson*

wet sand and pebbles that glistened under a cold clear sky. Not a tree in sight, not a shrub or flower, not even many lichens on the virgin boulders yet. Cloudy silt-laden streams and molten blue pools sparkled in the low spots, and remnant hill-sized blocks of decomposing ice hunkered down in the deeper hollows, sloughing off layers of dusty surface debris like old dogs shedding winter fur. Far off on the northern horizon lay an unfamiliar range of white, mile-high hills, the sun-scored southern face of the melting ice sheet. The vision lasted only a few moments, but a strong feeling of connection to long-ago times when Big Ice ruled this landscape stayed with me through the rest of my hike that day.

Let's continue with this imagination game. What if it happened again?

Here and now in the Adirondacks we worry, with good reason, about the effects of acid rain, invasive species, and global warming on our local ecosystems. But those problems won't exterminate every last Adirondack fish and fowl, and even the most extreme case of Anthropocene heating would still leave the land covered with some sort of greenery, if not all the kinds we're currently used to.

A full glacial advance, on the other hand, is a total wipeout. Every lake is bulldozed or smothered under a thick blanket of cobbles, sand, and gravel. Every sugar maple, every golden-tinted trout lily, every tuft of moss heaves up in bow waves of dirt and stones and is crushed to pulp. Every animal with legs or wings flees southward. The Adirondack peaks vanish under a heavy white tide, the iconic ski jumps at Lake Placid topple and are ground to splinters, and every settlement from Saranac Lake to Old Forge is obliterated.

Meanwhile, farther north, most of Canada disappears. That includes Quebec City, Montreal, Ottawa, Toronto, Winnipeg, Calgary, and Vancouver, not to mention every wild area from Hudson Bay to Banff. From a human perspective, there's no place called Canada for tens of thousands of years except in the same sense that a gigantic frosty slab called Antarctica now squats on the South Pole. And out across the Atlantic, advancing walls of white demolish Dublin, Liverpool, Oslo, Stockholm, Copenhagen, Helsinki, and Saint Petersburg, and every settlement on the rocky coastal rind of Greenland is shoveled into the sea by heavy spatulas of ice.

With much of the world's freshwater imprisoned in frozen form on the continents, sea level falls by as much as 400 vertical feet (120 m). The site of every twenty-first-century port is stranded far inland, the long, slender thumb of Florida doubles in width, and the present location of every shallow-water coral reef in the tropics sprouts weeds and trees. The associated cooling weakens monsoons, locking much of Africa and southern Asia into chronic droughts.

This is what a climate historian is likely to have in mind when discussing climatic change. Compare it to what most experts expect modern warming to bring us in the Anthropocene future and you'll understand why a paleoecologist's panic button might not be so easily pressed.

But wait. Isn't global warming supposed to trigger the next ice age? Isn't that what we saw happen in the apocalyptic enviro-thriller movie *The Day After Tomorrow*, in which the greenhouse effect suddenly shuts down climatically important ocean currents in the North Atlantic and triggers a superglaciation?

The movie isn't totally wrong, in that the warm Gulf Stream really does help to keep northwestern Europe from becoming cooler than it already is. It's part of a huge global conveyor belt system of interconnected currents that draws solar-heated tropical water into the cold surface of the North Atlantic, where it cools off and then sinks for a deep return journey southward. Some scientists worry that future climatic changes could disrupt that conveyor and trigger a sudden regional cooling; hence the movie scene in which a fierce wind seizes Manhattan with remorseless fangs of frost. But as gripping as that storyline is, serious questions remain about the real role of the conveyor in past and future climate change.

The engine driving the conveyor goes by several dry technical names, most recently the meridional overturning circulation, or MOC. It is also sometimes called THC, an abbreviation that is in no way connected to marijuana smoking (and tetrahydrocannabinol) but rather, reflects the upgrading of a simpler concept, that of thermohaline circulation, whose basic premise is that changes in temperature and saltiness drive major circulation currents of the oceans.

Warm water on the surfaces of the tropical oceans loses moisture to evaporation, which makes it saltier than average seawater. When the Gulf Stream flows from the hot latitudes between West Africa and the Caribbean into the cooler North Atlantic, it doesn't easily mix with those northern waters because its tropical heat content makes it less dense (warming makes both water and air expand). But the Gulf Stream gradually releases much of that heat into the cooler air over the North Atlantic, and when it finally does chill down its extra load of salt leaves it denser than usual.

That extra density makes some of the Gulf Stream water sink beneath the surface and continue its riverlike meanderings at greater depths. By the time it resurfaces, the deep flow has wormed its way around the southern tip of Africa and entered the Indian and Pacific oceans. Back on the surface again, the current recurves back across those oceans, rounds the tip of South Africa, and returns to the North Atlantic, picking up new loads of equatorial warmth along the way. Additional branches also operate in the Southern Ocean and Arabian Sea, adding extra loops to the tortuous path of the global conveyor.

There's a lot more to the picture than that, however, and when illustrations of this common version of the THC concept appear in professional slide presentations, they can become what one speaker at a recent meeting of the British Royal Society called "oceanographer detectors," because they make specialists in the audience "go visibly pale at the vast oversimplification."

The THC model is not so much wrong as incomplete. Most scientists have now switched the focus of ocean-climate discussions to the more comprehensive MOC formulation because temperature and salinity aren't the only drivers of ocean currents after all; winds and tides are at least as influential. THC-style flow does occur, but midlatitude westerly winds and tropical easterly trades do much of the actual pushing.

So why does marine MOC affect climate? As heat rises into the air from the Gulf Stream, it warms the westerly winds that blow toward Europe. Without those ocean-tempered winds, London might be as cold as . . . well, look at a map to see what lies at the same latitude on the opposite side of the Atlantic, and you'll find snowy Labrador.

With this basic introduction to the topic, you're already well enough equipped to take a pot shot at *The Day After Tomorrow*. The prevailing winds over Manhattan blow offshore toward the Atlantic, not from it, so why should a Gulf Stream shutdown freeze the city? The film also unrealistically subjects Europe to severe winter conditions year-round. Even if it really did become a climatic equivalent of Labrador, northern Europe would still warm up quite a bit in summer, just as Labrador does.

In reality, a MOC slowdown alone couldn't turn Europe into a climatic twin of Labrador because it lies downwind of a temperature-modulating ocean rather than the interior of a continent. And because prevailing winds spin the North Atlantic surface current system clockwise regardless of what the salinity or temperature of the water is, some version of the Gulf Stream will exist as long as these winds continue to blow over it.

Although some computer models do simulate moderate conveyor slowdowns in a warmer future, a truly severe disruption would require extremely large floods of freshwater to pour into the sea, presumably from the melting of land-based ice. If, say, a major ice sheet were to slide off into the North Atlantic where some critical sinking zone is operating, then perhaps it might cap the ocean off with dilute, buoyant meltwater.

In 1999, oceanographer Wallace Broecker published a striking theoretical description of just such a total MOC collapse under perfect-storm conditions. Tundra replaces Scandinavian forests. Ireland becomes the climatic equivalent of Spitsbergen, an island in the Norwegian Arctic. When climate modelers working at Britain's Hadley Center several years ago told their computers to "kill the MOC," the virtual air outside their lab cooled by 8°F (5°C) within ten years, at least on the digital screen.

But Broecker maintains that such a scenario is unlikely today, because those theoretical events only played out in a world that had already been cooled by a prolonged ice age. Nowadays, however, we don't have nearly enough readily meltable ice left in the Northern Hemisphere to do the job. To reset that stage we'd have to cover Canada, northern and central Europe, and Scandinavia with thick ice caps, and that would require colder, rather than warmer, conditions in the future.

Most computer models that have been upgraded so they more accurately represent the role of winds in ocean circulation foresee little, if any, cooling in the North Atlantic region from MOC disruptions during the Anthropocene. As the latest Intergovernmental Panel on Climate Change (IPCC) report concluded, "it is very unlikely that the MOC will undergo a large abrupt transition during the 21st century," and most experts believe that future greenhouse warming will overwhelm any minor regional effects related to MOC. In light of such findings, Broecker has tried to tamp down some of the worst exaggerations of the ocean-climate link that have been made by nonspecialists, but it's a tough struggle that pits scientific restraint against the lure of a good story.

One case in point is a study commissioned by the U.S. Department of Defense that presented a wildly extremist view of MOC collapse as a grave and imminent threat to national security. In their 2003 report, the authors noted that they were presenting only the most severe of all possibilities, as is commonly done in military planning circles, but that disclaimer was easily missed amid the frightening scenarios that followed. In their depictions, global average temperature shoots up faster and faster until, in 2010 AD, the MOC begins to collapse. Less than ten years later, according to their model, northern Europe cools by 5 to 6°F (3°C), devastating drought strikes the United States, and "a cold and hungry China peers jealously across the Russian and western borders at energy resources."

In response, Broecker wrote an open letter for publication in *Science* that expressed his dismay over the hyperbole. "I take serious issue with both the timing and the severity of the changes proposed," he wrote, pointing out that such extreme changes would take a long time to develop and would require glacial-type conditions, not global warming, to trigger them. Furthermore, he cautioned that computer models still can't fully reconstruct complex MOC disturbances of the past, much less those of the future. He concluded his letter with this admonition: "Exaggerated scenarios serve only to intensify the existing polarization over global warming."

Nonetheless, the idea of a total collapse of MOC is so emotionally gripping that it has become firmly lodged in public consciousness. In that context, oceanographers are watching closely for signs of conveyor responses to modern warming, just to be on the safe side. In 2005, for example, a team of British researchers described a 30 percent slowdown in MOC flow since 1957. The news caught fire among lay and professional audiences alike, but follow-up studies found that the slowdown alert was a "false alarm," as Richard Kerr put it in a deflating news brief for *Science*. The pattern of MOC flow is extremely variable, and a more careful look at the numbers showed that the reported trend was indistinguishable from random fluctuations.

If MOC changes are unlikely to freeze Europe, will ice ages play any role at all in an Anthropocene world? The answer is a qualified yes, but if they do reappear, it won't be the fault of ocean circulation disruptions. The main drivers of large-scale glaciation are the movements of Earth itself as it makes its annual elliptical dance around the sun.

The way many descriptions read in the popular media, you'd think that stopping our greenhouse gas emissions would prevent climatic change altogether. In fact, climates will always change whether we exist or not, just as they have even on Mars where similarly periodic frosts, thaws, and floods have left their signatures in deposits of red sand, gravel, and dust. Fortunately, because much of that change is cyclical, we can predict some of it by letting it play out on computer screens.

The fastest moves in the planetary dance are wobbles. Imagine a whirling top as it slows down and progressively loses its balance. The top begins to dip and swing around in progressively wider circles as it spins more slowly on its axis. Earth does that, too, though not because it's going to tumble over any time soon. In the wobble cycle, the North Pole draws a full loop roughly every 21,000 years (technically speaking, there are actually two modes of the cycle, 19,000 and 23,000 years long).

This has climatic effects because it changes how sunlight intensity, or insolation, affects different parts of the planet's curved surface. Each year, winter comes to the Northern Hemisphere when it leans away from the sun, and northern summer reigns when it leans toward the sun. Every 21,000 years or so, the wobble cycle brings the Northern Hemisphere its annual dose of summer only when we're farthest away from the sun on our egg-shaped orbit. When that happens, northern summers become slightly cooler than usual and less snow melts as a result.

Keep in mind here that these events don't happen because the sun itself changes. Instead, it's because sunlight affects the seasons and hemispheres differently over time. Those effects are amplified by geography because most major landmasses are crowded into the northern half of the globe and dry, solid ground accumulates ice sheets more readily than oceans do. For these reasons, ice ages typically begin in the Northern Hemisphere; the last one was born in the ice-friendly insolation target areas of northeastern Canada and northwestern Eurasia.

But that's not the whole story. Several longer cycles also influence the comings and goings of monster ice sheets.

As Earth wobbles, it also tilts more or less steeply in ways that amplify seasonal temperature differences. The slower tilt cycle takes 41,000 years for the North Pole to rock back and forth between 22.1 and 24.5 degrees of arc. When this cycle tilts the planetary axis less steeply, it aims each pole less directly at the sun during its seasonal summertime, so it warms less than usual. For reasons as yet unknown, this was the dominant pacemaker of ice age recurrences until about 1 million years ago, when the wobble cycle and a third, even slower cycle joined forces with the tilt cycle.

That third pulse, the eccentricity oscillation, changes the shape of Earth's orbital route around the sun. The path becomes more or less egglike over the course of about 100,000 years; it also changes in other ways every 412,000 years. Because the sun sits off center within that ring, snuggling a bit closer to one end of the oval than the other, Earth's distance from the sun varies a great deal through the seasonal circuit, and distortions of that route caused by the eccentricity cycle accentuate those changes. When the orbital ring is most distorted, a good deal less solar heat reaches us at the farthest end of the egg.

As all these cycles operate at once, they interact in ways that are easily visualized by comparing them to water waves. I first learned about this from my friend and colleague, ice core researcher Paul Mayewski, who directs the Climate Change Institute at the University of Maine in Orono. He explained how most of the ragged jumps and wiggles in his polar climate records originate.

"It's like waves on a lake," he began. I imagined orderly ranks of swells rolling along under a brisk breeze. "The rising and falling of the main swells are like the slow eccentricity cycle. Now imagine that a motorboat wake joins the pattern. Those waves are smaller and closer together, so they don't line up perfectly with the larger, wider ones."

I envisioned an irregular, bumpy surface like the ones that I often encountered during the waterskiing days of my youth. It was easy to keep my balance on a single predictable wave pattern, but when the driver doubled back on our trail or another boat dragged its own chop across our path, it spawned a crazed tangle of leaping, plunging waves. Where two crests briefly collided they bounced skyward, and where two troughs briefly met they bounced lower. Add forward and crosswise motions to such a collision zone and, as I can personally attest, a fallen water-skier will bob around in it like a cork.

"That's where a lot of long-term climate variability comes from," Mayewski continued. "When different cycles occur simultaneously, they harmonize and strengthen each other sometimes and they weaken or cancel each other at other times. And the more cycles you mix into the climate system, the more erratic it becomes." Mingling like waves through the ages, Earth's insolation cycles account for a surprising amount of natural climatic instability, and when they produce an exceptionally long and low temperature combination they can trigger a full-scale ice age.

These cyclic patterns were worked out by James Croll, a nineteenth-century Scottish scientist, and later refined early in the twentieth century by a Serbian civil engineer, Milutin Milankovitch. The fit between theory and history isn't perfect, and many of the past's shorter temperature perturbations had other causes. Furthermore, we still don't know exactly why northern glaciations have affected so much of the world at once; the slow gyrations of Earth tend to cool northern summers while warming southern ones, so you might expect ice ages to affect only half of the planet at a time. But the basic hypothesis of insolation-driven ice age pacing is still well supported in the geologic record. Take, for example, the history that was recently revealed by one superlong ice core from Antarctica. It represents 800,000 years of climatic change, and it also captures eight glacial cycles, reasonably in line with the 100,000-year rhythm of the eccentricity cycle.

Croll and Milankovitch worked with paper and pencil and devoted countless hours to hand-calculating past insolation values, a mind-numbing task that computers now repeat in seconds. But most interesting in the context of future Anthropocene climate are the models that run those cycles forward in time. By tracking Arctic insolation patterns into the future, we can predict when ice ages should come back to haunt us again—or, rather, when they would occur in the absence of our carbon pollution.

The wobble cycle is now trying to make northern summers slightly cooler than usual, which is one of the preconditions for launching a new ice age. And if we think of the last 11,700 years of postglacial conditions as just the latest in a series of similar interglacial warm spells, then common sense suggests that we're about due for a new cold snap.

During the 1960s, a temporary global cooling episode and a few scientists who incorrectly attributed it to the return of icehouse conditions launched a brief flurry of "fear the global cooling" excitement in the media. That response, of course, was misguided because the decadal time scale under consideration was far too short to represent Milankovitch-scale cycles, and what at the time may have felt like the start of a neoglaciation was in fact merely a brief pause in the twentieth-century warming trend.

But we can do much better than that today. Well-trained specialists now devote their careers to calculating the patterns and climatic effects of orbital cycles. Among the most influential of these are André Berger and Marie-France Loutre, a pair of climatologists at the Institute of Astronomy and Geophysics in Louvain-de-Neuve, Belgium. Their charts and tables of insolation data are widely cited in the scientific literature, and the graph shown here was drawn with data that were kindly provided by Dr. Loutre.

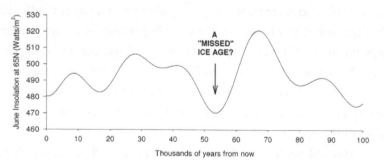

Solar insolation at 60 degrees North, showing an anticipated cooling event ca. 50,000 AD that might have triggered the next ice age in the absence of lingering fossil fuel emissions. *Data courtesy of Marie-France Loutre.*

So, when is the next ice age scheduled to arrive? A Berger-Loutre chart that appeared in *Science* in 2002 lays it out clearly. Today, the Arctic has entered a phase of relatively weak summer insolation that, in earlier times when the eccentricity cycle was stronger, might have been deep enough to call another ice age down upon us within the next few thousand years. But this episode will be too mild to do the job, even without our greenhouse gases working against it. In other words, we've just dodged an icy bullet, thanks to a slight, temporary diversion of our route around the sun.

Berger and Loutre's work also shows that the effects of the 412,000-year eccentricity cycle have been weakening in recent millennia, which helps to flatten out the depths of cooling pulses. The more symmetrical our orbit becomes as a result of changes in that longest of cycles, the narrower the range of seasonal temperatures on Earth. This will set an even calmer tone for cyclic climate changes in the more distant future of the Anthropocene.

Roughly 25,000 years from now, our orbit will be about as close to circular as it ever gets, and the power of the other cycles to warm or cool the ice masses of the Arctic will be as weak as it ever gets. Under those conditions, we'll still face brief regional disturbances, such as the North Atlantic Oscillation, El Niño, and the sheer orneriness of local weather that we know today, but they'll only raise short-lived climatic ripples on the flattened surface of the longer orbitally driven patterns. With the gentle urging of slightly higher axial tilt than we experience today, the curve of northern summer insolation will simply add a very mild rise to a puny peak—more like a low hilltop—about 25,000 years from now.

Although that low blip will cause only a mild thermal boost, it nevertheless does mean that Arctic summers were already slated to warm very slightly in coming millennia even without our influence. If Berger and Loutre are correct, then this may be unwelcome news to fans of low sea levels and heavy ice cover on Greenland.

According to the calendar of orbital cycles, the next serious risk of an ice-spawning chill during Arctic summer isn't due until about 50,000 AD, but here is where we humans enter the scene. Most climate models will only trigger an ice age at that distant point in future time if the CO_2 content of the atmosphere is no higher than 250 parts per million (ppm).

With CO_2 concentrations as of this writing at 387 ppm and rising, they clearly won't be going back below that critical threshold for a very long time. In fact, as will be explained in Chapter 3, those values will still hang well above 250 ppm in 50,000 AD, and they won't return to such preindustrial conditions for tens of thousands of years beyond that.

These two factors together—the inherent weakness of the next cooling cycle and the longevity of our carbon pollution—lead to one astounding conclusion. Not only have we warmed the world during this century with our carbon emissions; we've also stopped the next ice age in its tracks.

According to recent computer model studies, this may be only the beginning of our influence over the coming and going of ice ages, depending on what we do during the next century or two. If we hold ourselves to a relatively moderate sum total of fossil fuel emissions before switching to alternative energy sources, then we skip the next glaciation in 50,000 AD. But more of those natural cooling pulses will develop later on. When the next big one strikes the Northern Hemisphere some time around 130,000 AD, the greenhouse pollution legacy of a moderate-emissions scenario will have faded to insignificance, leaving nothing to resist glaciation. Once again, ice will bury the northern sectors of the northern continents. Or rather, it will do so unless we burn through all of our remaining coal reserves during the next century or so. If we go down that profligate path instead, there won't be any ice ages for a very long time: not in 50,000 AD, not in 130,000 AD, and not during the other insolation minima of the next half million years.

How should we react to this news? When I imagine our carbon footprint kicking the door closed on an ice age, my initial reaction is one of shock mingled with a twinge of fear. But a second, rather confusing response follows when I imagine our carbon emissions saving the northern United States, Canada, and much of northern Eurasia from being crushed under gigantic ice sheets.

It would be stretching the bounds of credibility to say that greenhouse pollution is anything but a vexing problem. But on the scale of planetary disaster, an ice age is to global warming as thermonuclear war is to a bar brawl. We're certainly justified in worrying about the environmental disturbances that are coming our way, but even this situation is arguably better than losing entire nations and ecosystems to total icy destruction. What if we had to choose between the two options?

In fact, we do face that very choice as we weigh our possible responses to modern climate change. Most of the more extreme negative effects of today's fossil fuel emissions will be felt within the next millennium or so, but what if those same fumes later go on to save much later generations from having to endure otherwise inevitable ice ages?

At first, this ethical dilemma might seem outlandish, a silly kind of joke. Speaking of it too flippantly feels disrespectful of those who may suffer from the climatic changes of today and in the near future. It also feels like tossing raw meat to the habitual contrarians who seek any excuse to avoid controlling fossil fuel consumption. But the facts are plain, and I believe they're worth considering carefully.

Unfortunately, the large time scales involved in this subject complicate discussions of it. It's much easier to imagine what climate change might do to ourselves or our grandkids than what it might do to people 130,000 years from now, and this can make it difficult for some individuals to take the matter seriously. For them, weighing the value of their own lifetimes against those of unknown citizens of the far future is as ridiculous as weighing the value of a living child against that of a cartoon character; such distant people and times are simply too remote to seem real.

On the other hand, many of us do enter that deep pool of inquiry when we consider doing things for the sake of future generations, though our entrance is usually more of a toe dabble than a dive. In fact, thinking and acting on behalf of future-dwellers isn't the simple endeavor that it may seem to be, especially if you don't restrict your attention to those closest to us in space and time.

For one thing, you have to decide who those beneficiaries will be. Does your list include everybody who will be alive in, say, 2500 AD or just a select subset of them? You might be willing to give up your gas-guzzling car in order to keep the climate of a direct descendant's world similar to that of your own. But what if that choice also harms someone else, perhaps by making some future nation much wetter or drier than it would otherwise be: will you favor your own lineage over those of others? And if you choose to favor your own descendants, you still have to decide which ones to focus on because various generations will prefer different things at different places and times in the future.

If we imagine the Anthropocene in its entirety, it gives us a very large and diverse collection of future people to consider. Many of them will inhabit a warming world, but most will live on the long cooling tail-off of atmospheric CO_2 recovery. Perhaps citizens of 130,000 AD might decide that, after so many millennia of ice-free oceans and high-latitude commerce, they'd rather keep things the way they are rather than let CO_2 concentrations return to preindustrial levels, particularly with a new ice age looming. Maybe by then our long-lived carbon emissions will have begun to seem less like pollution and more like insurance against global cooling.

Then again, maybe not. But who are we to decide their fate, anyway?

We will be messing with planetary temperatures for a very long time, well past the temporal limit of 2100 AD, but we still have time to head off the most extreme consequences if we choose to do so with a long-term view of the future in mind. One of the most important decisions that our generation will ever make is the critical choice between a relatively moderate scenario in which we quickly replace fossil fuels with other energy sources and an extreme one in which we burn through our remaining cheap petroleum and then go on to consume our coal reserves as well. But in either case, one thing is becoming increasingly clear. Humankind is going to face a long-lasting array of environmental challenges of its own making as the Anthropocene runs its course.

2

Beyond Global Warming

Carbon is forever.

—*Mason Inman,* Nature Reports Climate Change,
November 20, 2008

Fossil carbon threads a common causal link through the Anthropocene from long-term warming to sea-level rise, and it has already made some enormous environmental changes inevitable. But the details of exactly how they could play out are still largely up to us. As luck would have it, we who live in the twenty-first-century will be powerful decision makers in this new Age of Humans. Before us lie several possible routes, and we're going to select one of them on the basis of how much fossil fuel we choose to burn during the next hundred years or so. As we struggle to make that choice wisely in the midst of our self-made carbon crisis, we need to understand as much as possible about the ways in which our actions today may influence the long-term future of the world.

In these early days of human-driven climate change, global warming is on center stage. But that is only the foreword to an immensely long tale in which climate change will mostly come to mean global *cooling*, albeit from artificially boosted temperatures that will remain higher than those of today for thousands of years. According to an increasingly influential group of climate visionaries whom I'll introduce shortly, the slow natural processes that must gradually erase our carbon footprints will take not just a few more centuries but many millennia to lower CO_2 concentrations to their normal ranges. As they drop, artificially high greenhouse temperatures will also fall in response.

But those changes will occur far more slowly than most of us yet realize. A full return to preindustrial conditions will take tens of thousands—perhaps even hundreds of thousands—of years.

It's not difficult to visualize the basic trajectory of things to come if you keep in mind the following maxim: What goes up must come down. I will assume here that you already know how increasing concentrations of greenhouse gases such as carbon dioxide (mainly from power plants, vehicles, and factories) and methane (farms, landfills, sewage, mines, pipelines) raise temperatures by trapping solar energy in the atmosphere as it re-radiates from Earth's sunlit surface in the form of heat; detailed descriptions of that process are now ubiquitous online and in print. But greenhouse gas concentrations in the air can't increase forever because there are only a finite number of accessible carbon atoms in the world. And temperatures can't rise indefinitely, either; even the worst possible scenario won't make the planet burst into flames. An eventual peak and turnaround in the modern warming trend are therefore inevitable.

Most of us are still so focused on the present pattern of warming that it can be easy to forget the obvious, that entirely different modes of change must lie ahead. At some point, the direction of current trends will shift into reverse and produce a staggered climate whiplash effect. Coping strategies that have been developed in response to warming will become obsolete, or even liabilities, as the adaptive momentum gained in adjusting to rising trends pushes onward while the environmental setting itself lurches off in new directions. As future temperatures pivot into cooling mode, not only will Earth's inhabitants have to adjust to environmental change; they'll have to adjust to a complete reversal in the very nature of that change.

But it won't be quite as simple as a single episode of whiplash might be. The loose connections among various components of the air-land-ocean system will cause delayed reactions to the changes driven by the rise and fall of greenhouse gas concentrations. Those components will act like a string of mountaineers loosely roped together; if the lead climber slips from an icy crag, the remaining team members will be pulled one by one after intermittent moments of slack time.

You need no computer program to predict the basics of this coming stage of global change; deducing its general outline is largely a matter of common sense, though the specifics of timing and intensity are another matter. Sooner or later, we'll stop emitting so many greenhouse gases, either by choice or by running out of affordable fossil fuels. After our emissions rate peaks and begins to decline, the atmospheric CO_2 curve will also crest and then begin to fall back toward preindustrial levels. All else will follow the lead of that CO_2 pulse.

The next wave of whiplash will come as global average temperatures peak at some thermal maximum and then drop back down in a delayed reaction to the CO_2 reversal. The very idea of an end to global warming may come as a surprise if you have limited your thinking on this topic to the short term, but it simply can't go on forever. The greenhouse effect will inevitably weaken in response to the future decline of fossil fuel pollution, and that will bring temperatures back down.

Under the cooling-but-still-warm conditions that follow the thermal peak, polar ice sheets will continue to shrink and the heated oceans will continue to swell in their basins like yeasty dough in a warm kitchen. But eventually, the rising seas will reach some peak elevation and then begin to fall along with temperature. So, too, will the acidity levels of the oceans.

To examine that bumpy road ahead more closely, we must look far beyond 2100 AD into the realm of reasonable speculation. And to put approximate dates and magnitudes on these coming waves of change, we can now be guided by pioneering scientists who probe the future with new breeds of global climate models.

But how can we trust such models if forecasters can't even predict local weather just a few weeks in advance? True, computer models are to real climate as model airplanes are to fighter jets, but we aren't trying to make far-future predictions about weather, which is a short-term, localized, and rather chaotic phenomenon. Instead, we're talking about basic, long-term, global-scale climate, which is averaged over many years and over large areas of the planet. We're only asking for broad generalizations here, and ones that are based upon sound scientific principles. For example, no responsible climatologist would seriously proclaim that July 29, 5000 AD in New York City will be a bright and sunny day, but we can be reasonably sure that dawn will arrive from the east on that morning and that water will neither freeze nor boil under the conditions that July would be likely to bring to that location.

A key concept underlying these projections is the conservation of matter. When we shut down the spigot of air pollution, our emissions won't simply vanish. They may drift downwind, but they still exist somewhere on this bubble of a planet. We may watch chimney smoke dissipate, and we speak of throwing trash "away," but the atoms in that smoke and that garbage are neither created nor destroyed by our normal daily behavior, even though we may have forgotten about them.

Skilled modelers can follow the wanderings of our carbon emissions in computer-generated worlds like trackers on a fugitive's trail. After all, there are only so many places for a carbon atom to go. It can drift freely in the air for many years, but it will eventually take up residence elsewhere. It might dive into the sea before reentering the air on the other side of the world. It might be sucked into the tiny pores of an oak leaf and spend a tree's lifetime bound into the structures of bark or wood. Later still, it might dissolve into a raindrop that splashes against a granite boulder and works its way into the crystal lattice of a feldspar grain, helping it to crumble into dirt. And today it might be lodged in the fleshy tip of your nose, anticipating its next move to some other earthly destination. The best computer models take this shape-shifting into account as they probe the future.

One of the earliest and most prolific sources of information on this topic is David Archer, a climate-savvy oceanographer at the University of Chicago with the professional resumé of someone with too much inquisitive energy to fit into a single lifetime's work schedule. Using a new generation of sophisticated computer models with names like CLIMBER (Climate and Biosphere), GENIE (Grid-Enabled Integrated Earth System), and LOVECLIM (a composite of five other model names), Archer and a growing corps of like-minded investigators around the world are tracing and refining predictive templates for a future that is flooded with the vapors of combusted coal, oil, and natural gas.

"The idea that a sizable fraction of our carbon dioxide could stick around for hundreds of thousands of years hasn't reached mainstream consciousness yet," he told me in a recent phone conversation. And that goes for most of the scientific community as well, although the idea is percolating in ever wider circles these days. This field of inquiry is so new that you can still count most of its foundational publications on your fingers. Kirsten Zickfeld, a climatologist and computer modeler at the University of Victoria, British Columbia, acknowledges Archer's ground-breaking role in the field but adds one important reason why this kind of research was not being conducted earlier. "We simply didn't have the right tools for it until just a few years ago," she explained to me. "The models have to mimic the cycling of carbon in and out of various habitats along with the climatic changes. Older simulation systems weren't fast enough to handle so much complexity." These upgraded models, in turn, are rapidly upgrading the time scales that we imagine the future on. "We can now see the irreversibility of the impacts that we're having on the planet today," she continued. "Our carbon emissions won't be gone nearly as quickly as we once thought. In fact, they'll stick around pretty much forever."

Simulations of our near-term carbon future present such a diverse collection of twenty-first-century trajectories that it can be difficult to know which ones to focus on as we hitch them to the long-term outlooks that Archer, Zickfeld, and their associates are generating. Rather than try to sort through all of them here, I've selected a representative pair of carbon emissions scenarios from the last assessment report of the Inter-governmental Panel on Climate Change (IPCC), one of them relatively moderate and one extreme, to bracket the most commonly accepted range of paths that lie before us. They will provide the basics of what you need to know about our main options for the near future. I'll then offer much longer-term views of what could follow these two opening acts, based upon the outputs of CLIMBER and other models. The results of those simulations vary somewhat in their specifics, but in general they are remarkably consistent.

In the moderate scenario, we aggressively limit CO_2 concentrations to a peak of 550 to 600 ppm. The IPCC calls this low-growth emissions scenario B1. The B1 scenario is "moderate" only in relation to the more extreme one that follows, and many climate activists are currently focused on holding greenhouse gas concentrations to a much lower 350 ppm, a level which is considered likely to prevent societally disruptive human-driven climatic changes. I support and am inspired by 350.org and related movements, but we are already well past that preferred milepost on the rising carbon curve, and most of the climatologists I have spoken to believe that we won't be going back to those concentrations any time soon. The choice of B1 for the low end of the scenario spectrum is based more on a sense of hard-core realism than on desire. The 350 ppm concentration is an excellent goal to aim for, but 550 to 600 ppm is probably what we'd really end up with after taking such aim.

In this so-called moderate case, we switch to nonfossil fuels as soon as possible but still end up adding another 700 gigatons (Gton; 1 billion metric tons) of carbon pollution to the 300 Gtons we have already released since the Industrial Revolution, bringing us to a grand total of 1,000 Gtons. Although the environmental changes that are likely to follow such a release will be large and long-lasting, and even though it would be preferable to avoid such a path altogether, this hypothetical situation is meant to approximate a realistic best-case—or rather, least-unwelcome—scenario for our Anthropocene future.

Scenario 1: A Moderate Path

We can't realistically expect fossil fuel consumption to stop everywhere all at once, so let's say that our rate of CO_2 emissions reaches its peak around 2050 AD and finally fades out altogether by 2200 AD.

In that case, atmospheric CO_2 concentrations climb from today's value of 387 ppm to a peak of 550 to 600 ppm, roughly twice the preindustrial level of the 1700s, some time between 2100 and 2200 AD. The oceans gradually absorb much of that excess CO_2, which becomes carbonic acid in solution and changes the chemistry of seawater, making it increasingly corrosive to the limy shells of many marine organisms, especially in colder waters at high latitudes and great depths.

When thermal maximum arrives around 2200 to 2300 AD, global mean temperature may be 3 to 7°F (2 to 4°C) higher than it is today. That peak could come a century or more later than the CO_2 peak because of what climatologist Tom Wigley calls "climate change commitment," a delaying action that mainly reflects the slow response of oceans to heating. Even if we could halt our greenhouse emissions altogether right now, we would still have to face a degree or more of additional warming over the next century. Taking this kind of delay to the extreme, some computer simulations of our moderate scenario by University of Victoria researcher Michael Eby and colleagues push the date of the thermal maximum at least 550 years after the CO_2 peak.

As the temperature peak passes, the whiplash effect ushers in a prolonged global cooling-off period. But even though the direction of change has flipped from warming to cooling, the world is still hotter on average than it is today and polar ice continues to melt and flow into the oceans, lifting the sea surface by several feet per century. In addition, the deeper, colder layers of the oceans are still warming and expanding. Thermal expansion in this case could drive global mean sea level an additional 1 to 2 feet (0.5 m) upward, but how fast and far it eventually rises before experiencing its own whiplash reversal will mainly be determined by how much land-based ice melts. In this relatively moderate scenario, one might reasonably expect to lose about half of Greenland's ice and much of the West Antarctic ice sheet while leaving the huge East Antarctic ice sheet largely intact. As a result, sea-level rise could finally stall at 20 to 23 feet (6 to 7 m) above today's elevation several centuries or millennia from now.

So far, none of this seems particularly surprising, being standard fare for current descriptions of global warming. But it's at this point in the climate narrative that Archer and other investigators have begun to grab the scientific world by the collar. Although many who discuss the duration of global warming still limit it to a few hundred years, a growing body of new evidence shows that a century-sized time scale for full climate recovery is far too short to be realistic. "The lifetime of fossil carbon dioxide in the atmosphere is several centuries . . . plus 25 percent or so that lasts essentially forever," says Archer.

He reached that conclusion by considering where our massive slug of CO_2 may go once we stop pumping additional fumes into the air. By burning fossil fuels, we're emptying a global graveyard of plants and plankton that took millions of years to fill, and there'll be no return to those tombs for many thousands of years. Archer's reference to "forever" is not meant to be taken literally; any carbon imbalance that we impose on the atmosphere will eventually level out again. But it will only happen over time periods that when fully grasped, can take one's breath away. From the frame of reference of any single civilization, say, ancient Egypt or the Roman Empire or the modern Industrial Age, the span of global carbon recovery time might as well be forever.

This may sound like a wildly extremist claim, but it most certainly is not. If you follow the reasoning behind it step by step, you'll find that it makes perfect sense. Just keep in mind that the carbon in our CO_2 and methane emissions doesn't leave the planet and it doesn't break down; carbon atoms are virtually immortal (except for the radioactive isotope carbon-14, which I'll discuss later). They have to go somewhere when they are released from a burning bit of fossil fuel, and they may wander from place to place after leaving a particular smokestack, flue, or tailpipe.

The computer projections show that, at first, much of our emitted carbon will dissolve in the sea. Gases enter water bodies quite easily from the stirring action of surf and currents as well as the diffusion of air molecules directly into liquid surfaces; that, after all, is how fish manage to breathe dissolved oxygen while swimming in a lake or ocean. The oceans actually contain about fifty times more CO_2 than the atmosphere and cover 70 percent of Earth's surface, and it is this uptake process that will drive the whiplash reversal of CO_2 concentrations after our emissions begin to trail off. But there's a limit to how much gas even an ocean-sized reservoir can hold.

After a millennium or two, the aquatic uptake of our carbon emissions will have slowed to a crawl. When that happens, sometime around 3000 to 4000 AD, between a fifth and a quarter of our CO_2 pollution could still be drifting about in the air. In the case of our moderate-emissions scenario, that would represent enough excess greenhouse gas to keep global mean temperatures 2 to 4 degrees (1 to 2°C) higher than they are now, even after a long recovery from the earlier thermal peak.

If Earth were made entirely of water, the gas satiation of the oceans would mark the end of the recovery process and a greatly reduced airborne remnant of extra CO_2 would be marooned above the waves eternally. But much of the planetary surface consists of dry land, and geological features underlie the watery regions, as well. These features, the rocks and sediments of Earth's crust, will clean up whatever remains of our pollution legacy. Unfortunately, they work slowly. Very, very slowly indeed.

The fastest repair mechanisms in the geological tool kit, or rather the least sluggish ones, involve chemical reactions with lime-rich carbonate materials such as limestone, chalk, and the shells of marine organisms.

These are the kinds of alkaline substances that fizz when you drop acid on them, like the baking soda that you may have dribbled vinegar on in science class so you could watch it erupt into foam. Carbonate rocks and sediments that lie exposed to the elements can be attacked by natural acids in precipitation, soils, and water, most importantly by CO_2-derived carbonic acid that now increasingly contaminates raindrops and water bodies worldwide.

Some of the first of such cleanup mechanisms to kick into gear will be those that lie in submerged marine environments. As our CO_2 pollution dissolves into the oceans during the early stages of the Anthropocene future, it will acidify them enough to dissolve some of the alkaline muds, corals, and shells that litter the bottoms of the ocean basins. As you might imagine, this will not be a welcome change for many sea creatures—but there is a positive side to it as well. As carbonate molecules move from those solid forms into solution, they will help to neutralize the acidification over the course of millennia, and by readjusting the chemical balance of the oceans in this manner they will also help them to absorb even more CO_2 from the air. In terms of the long-range global cleanup, it will be like feeding antacid pills to the ocean so it can continue to feast a bit more on the carbon pollution banquet.

Meanwhile, the carbonate rocks and soils on land will also be pitching in, but they will be dealing with rainwater, not seawater. You may be familiar with industrial acid rain, the strong stuff that kills forests in Europe and sterilizes lakes in my home region of the Adirondacks. Natural carbonic acid rain is different: in mild doses, carbonic acid is just another normal component of rainwater, and much weaker than the sulfuric and nitric acids that coal-fired power plants and automobile engines produce. It comes from the diffusion of CO_2 from the air into the water droplets in clouds, and it makes even the purest rain slightly acidic. When carbonic acid rain falls on, say, a pale lump of limestone, you won't see the rock foam up in response. But if you could look much closer, down on the microscopic scale, you'd see the wetted spots on the stony surface begin to crumble ever so slightly.

Imagine that we're watching this take place on some beautiful geological formation in a region where limestone is abundant, as in the jagged, spectacularly etched landscape near Guilin and Yangzhou in southern China, that is commonly depicted in traditional watercolors. Perhaps you've seen such paintings in which the steep-sided hills resemble conical towers or canine teeth, seemingly too tall, narrow, and sharply pointed to be real. Those dramatic shapes were sculpted over the ages by the erosive effects of naturally acidic water on soft, soluble rock through processes that also shape other limestone-rich landscapes around the world.

Let's say that we're standing on top of one such spire overlooking a bend in the Li River on a gray, misty day, and it's starting to drizzle. Small puddles pool around our feet, and they funnel their slightly acidic contents into flowing cracks and crevices where carbonic acid molecules tear at the crystal lattices of the limestone. Those reactive acids then morph into new, less corrosive molecules, called bicarbonate, each of which still carries its original atmospheric carbon atom with it, and tumble downhill with other substances that have fallen away from the slowly decaying stone and dissolved into the watery runoff.

The bicarbonate-enriched rivulets pour their contents into the river, which eventually delivers its load to the South China Sea. When the bicarbonates join the immensity of the ocean, they help to neutralize the marine acid buildup just as the seafloor carbonates did, and their additional dose of natural antacid helps the seawater to take another few bites of CO_2 from the air. Over time, bicarbonates and carbonates may also drag their carbon atoms down with them for more permanent burial on the ocean floor because marine organisms use such substances for photosynthesis and the building of shells and skeletons, turning them into heavy ballast when the host organisms die.

The upshot of this seaward migration of carbon atoms, which happens day and night, year after year, is that carbonate rocks and sediments on land will help to remove excess carbon from the air and store it in the oceans in much the same manner as the marine deposits did. After about five millennia of combined cleanup efforts, by which time the worldwide supply of carbonate materials has done about as much as it can, only 10 to 20 percent of our carbon excess remains in the atmosphere. But even so, that's still enough to hold global average temperatures 2 to 4°F (1 to 2°C) higher than now. Far out in 7000 AD, the lingering gases still hold enough greenhouse warmth to keep leftover polar ice melting and sea levels rising.

But even this is not the end of the story. Amazingly, 7000 AD is only the entry ramp to what Archer calls "the long tail of the CO_2 curve." Our atmospheric carbon pollution will continue to decrease, but the remaining recovery will proceed so slowly that another 5,000 years will barely dent the carbon residue. In 12,000 AD—ten millennia after our fossil fuel emissions have ended—global average temperatures could still be at least 2°F (1.1°C) warmer than today.

The waning tail of the carbon curve is so long because, once the oceans and carbonates do what they can, the last major cleanup mechanism left will be the hard crystalline frameworks of more resistant silicate rocks such as pink or white granite and dense black basalt. Their response to weathering by acidified water is somewhat similar to that of carbonates, but it is achingly slow. Just how long will it take to return to our present CO_2 concentration of 387 ppm by this route? Most computer models show that similar concentrations won't be seen again for at least 50,000 years, and complete recovery could take several times longer still.

Although the world will be cooling slightly throughout the long tail-off, those falling temperatures will still be higher than they would have been without our influence. That will allow plenty of time for polar ice sheets to lurch and dribble into the sea, potentially raising ocean levels by tens of feet after many millennia of mild but sustained heating.

What Archer and researchers like him are telling us here is that even the emissions scenarios that many investigators call moderate are locking us into immensely long chains of environmental consequences. But this example of a relatively mild carbon pollution scenario pales in comparison to the more extreme side of the emissions spectrum. What if we throw caution and everything else flammable to the winds and spew a whopping total of 5,000 Gtons of carbon into the air? This would fall at the extreme upper end of the IPCC list of possible futures, something like the fossil fuel–intensive A2 emissions scenario carried well beyond 2100 AD. It is often said that we're undertaking a huge ecological experiment by running our civilization on carbon-based fuels, but if the preceding 1,000-Gton scenario is just an experiment, then an extreme 5,000-Gton scenario is more like a crash test. If we go down that route, most likely by continuing on our current path until we burn through what remains of our coal reserves, then we can rightly be called crash-test dummies.

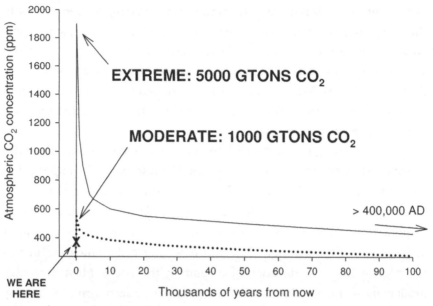

Atmospheric carbon dioxide concentrations for the next 100,000 years under two emissions scenarios. Recovery from the more extreme scenario takes much longer than 100,000 years. *After Archer, 2005*

Scenario 2: A Super-Greenhouse

Because it takes a long time for us to consume all of our easily accessible coal, CO_2 emissions reach their peak later than in the previous scenario, say, between 2100 AD and 2150 AD, and they continue at decreasing rates for another couple of centuries.

Atmospheric CO_2 concentrations in most computer simulations peak near 1900 to 2000 ppm, five times higher than today, around 2300 AD before reversing into a long-term decline. The carbonic acid spawned by all that CO_2 dissolving into the oceans becomes a corrosive solvent to shell-bearing sea creatures from pole to pole.

Temperatures peak as early as 2500 AD or as late as 3500 AD, depending on the model and the environmental parameters used; most simulations stretch the broad thermal maximum out over several centuries. The massive peak is so flattened and long-lived that it can be difficult to call it a point of climatic whiplash; it's more of an all-consuming high tide than a steeply cresting wave. Global mean temperatures during that thermal maximum could be at least 9 to 16°F (5 to 9°C) higher than today, or more than twice as high as the peak in the more moderate scenario. And this range merely represents the worldwide average; northern high latitudes might warm twice as much. Under those conditions, Europe, Scandinavia, and most of the United States are largely snow-free in winter.

By 4000 AD, many models drop atmospheric CO_2 concentrations only as low as 1,000 to 1,300 ppm or so, still about three times higher than those of today. However, a model described by oceanographer Andreas Schmittner and colleagues in *Global Biogeochemical Cycles* keeps those concentrations even higher, more like 1,700 ppm, a level close to that obtained by Michael Eby's team, as well. In that situation, sea-surface temperatures are also much warmer than they are now, perhaps 11 to 13°F (6 to 7°C) warmer in the tropics and 18°F (10°C) at higher latitudes. Such a global hothouse would sharply reduce the climatic differences between latitudes and leave the world more climatically homogeneous.

In 7000 AD, sluggish silicates continue to pick at the carbon left-overs in the atmosphere. One paper published by Archer and German climatologist Victor Brovkin suggests a cooling recovery of less than 2°F (ca. 1°C) during those first five millennia, after which CO_2 concentrations still hang close to 1,000 to 1,100 ppm. Even 10,000 years down the time line, between a tenth and a quarter of our greenhouse pollution still

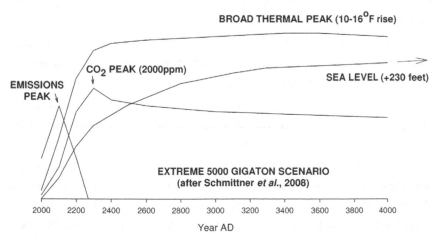

Detail of near-term changes expected for an extreme 5,000 Gton emissions scenario. *After Schmittner et al., 2008*

hangs in the air, keeping global average temperatures 5 to 11°F (3 to 6°C) warmer than those of today. Carbon dioxide concentrations don't resemble those of today until 100,000 AD or so, and full recovery in this extreme scenario takes at least 400,000 to 500,000 years.

With such prolonged and intense heating, Greenland's ice sheet eventually shrinks to bedrock and drains enough meltwater into the North Atlantic to lift sea levels by 23 feet (7 m). The loss of even more ice from Antarctica eventually pushes those levels ten times higher, gradually turning low coastal plains into submarine extensions of the continental shelves.

The differences between the moderate 1,000-Gton path and the devil-may-care 5,000-Gton route are profound and disturbing. In both cases, most of the environmental changes that they cause will occur slowly in comparison to our lifespans, but it's their magnitude and duration, more than their speed, that boggle the mind, particularly in the extreme case.

In the moderate scenario, most of the largest environmental disturbances are over and done with during the first millennium or two, though a small fraction of our carbon pollution lingers for tens of thousands of years. But an extreme 5,000-Gton emissions release lifts the long tail of the CO_2 curve much higher and stretches it out much further into the future, not just because our emissions were greater in that case but also because the warming itself launches feedback mechanisms that release natural stores of carbon and boost temperatures even further. The thermal maximum in this scenario is at least twice as high as in the more moderate case and it ends several centuries later.

Believe it or not, this sketch of an extreme hothouse future is actually conservative. One simulation published by oceanographer James Zachos and colleagues holds CO_2 concentrations nearly flat within the 500 to 600 ppm range to the end of the 100,000-year model run. And another study puts the end of the silicate cleanup a full million years in the future, even for a moderate 1,000-Gton scenario.

As we contemplate the expansive time scales that these changes will play out on, it can be difficult to believe that they are real. Looking so deeply into the future can be like sailing off the edge of a flat world in our imaginations, and for many of us nothing that is concrete and familiar lies beyond the mysterious horizon of our own lifetimes. Will people even exist in 100,000 AD?

When I've presented this question to my students and colleagues, most of them have said that they expect the human race to be extinct long before that date. Some expect a natural or human-made disaster to kill us off, and others believe in various religion-based dooms for humanity. I find this fatalism disturbing, because our survival as a species is central to all discussions of fossil fuel consumption, climate, and life on Earth. If we're not going to be here for much longer, then who cares how long our CO_2 will hang around in the air? And if there's not much time left for our descendants to live in, then why shouldn't we simply use up our natural resources today and enjoy life selfishly while we can?

Claiming that humans will not exist in the deep future can also be a cop-out. Eliminating a long-term future makes it easier to ignore the immense longevity of our carbon pollution and pretend that nobody will have to deal with the environmental changes that we're setting in motion today. But there's no way out from under this yoke of responsibility, because humankind is really not going anywhere. The lifestyle decisions that we make today will affect many generations to come, and the least we can do on their behalf is to acknowledge that fact.

The list of true threats to human existence is remarkably short, and most of the items that I've heard people put on that list are easy to discount. Industrial toxins are too diffuse or localized to drive us all to extinction. Volcanoes are out, too; even the most violent eruption in the history of humankind, the Toba supervolcano that exploded in Sumatra nearly 75,000 years ago, didn't kill everybody. And intense gamma ray bursts from dying stars or newly formed black holes are too rare to represent a credible threat to our species; even if such a burst did strike us, the broad hulk of Earth would shield the residents on the other side, so the damage wouldn't be total.

What about pestilence? One of the worst microbial killers of all time, the Black Death, killed between 75 and 200 million people during the fourteenth century, including nearly half of all Chinese, at least a third of all Europeans and Middle Easterners, and perhaps an eighth of all Africans. Nonetheless, it wasn't even close to the End of Days. Most Chinese, Europeans, Middle Easterners, and Africans survived, even in the absence of modern treatment, which today would drop the mortality rate to 5 or 10 percent.

What helps humankind to resist total extinction by disease? It's a process that many people don't even believe in: evolution by natural selection. Given the large numbers of humans in the world and the ubiquity and diversity of genetic mutations, there will probably always be some individuals among us who have built-in immunity to any given microbial disease. The more of us there are, the greater the genetic diversity and the greater the chances of someone being naturally immunized against infection. This is how life adapted to, and eventually became dependent upon, global oxygen pollution during the last 2 billion years, and even the worst pandemics can be seen as natural-selection events that favor the survival of resistant variants.

Will we wipe each other out with a thermonuclear war, thereby avoiding the trials of future climate change? Even more destructive than the fireballs and radiation would be the dust and smoke-induced nuclear winter that would leave a miserable remainder of humanity grubbing for sustenance. But this planet is huge, there are billions of highly resourceful humans living on it, and radioactive fallout spreads unevenly on the winds and sinks away under soils and ocean muds. Only a full-on, suicidal conflagration among multiple superpowers could kill every last one of us, and although it is an unpleasant possibility, I believe (with fingers crossed) that it is unlikely.

And finally, we do have a potential asteroid problem in this rock-filled solar system of ours. The lightning-fast chunk of space debris that blasted a 110-mile-wide (180 km) crater into Mexico's Yucatán Peninsula 65 million years ago might have decimated the dinosaurs. Steam and dust from the impact choked the atmosphere, and the heat of vaporizing rock ignited wildfires over thousands of square miles. But it would take a much more intense collision than that one to kill us all. The impactor would probably have to be large and fast enough to melt much of Earth's surface or break the whole planet into pieces. For those self-tormenting readers who might actually enjoy wallowing in fearful anticipation of asteroid disaster, it's worth experimenting with the "catastrophe calculator" that is posted online by the University of Arizona's Impact Effects Program. The website lets you enter information about the size, speed, density, and angle of the flying object as well as the nature of the target and how far away you happen to be from the impact site. With the click of a button, the calculator produces detailed and darkly fascinating descriptions of what happens next.

Unable to resist the temptation myself, I recently found that if an asteroid 1,000 feet wide (ca. 300 m) hits the ground thirty miles away from my home at 10 to 11 miles (17 km) per second, it will blast a hole 3.5 miles (5.6 km) across and more than a mile deep into the solid rock. A fireball twenty times brighter than the sun will fill the sky and ignite the forest around me, and an earthquake measuring 6.9 on the Richter scale will shake the ground. Then, a minute and a half later, jagged chunks of stone the size of my head will begin to crash down around me. Half a minute later still, a raging blast of hot wind will flatten the burning trees, my house, and me with a deafening roar. Not a total planetary wipeout, by any means, but it would certainly distract an observer from global warming for a while. To melt the whole surface of the planet, the projectile has to be at least 4,350 miles (7,000 km) across; anything much larger will shatter Earth into a spray of new asteroids. But the calculator puts a damper on the shock value of that result by noting that rogue impactors of this magnitude don't exist in our sector of the galaxy.

This review of options for a future apocalypse raises another important point worth keeping in mind as we gauge the severity of risks we face from our two carbon emissions scenarios. The environmental changes that may accompany them are too important to ignore, but they're not worthy of utter panic or despair, either. If asteroids or other truly serious threats aren't likely to destroy the human race within the next 100,000 years, then the greenhouse effect certainly isn't going to either. In order for it to wipe us all out, it would have to overwhelm more than 6 billion members of an incredibly resilient species that has thrived for tens of thousands of years in every imaginable habitat from frigid pole to torrid desert, even without modern technology and despite abrupt swings between glacial and interglacial conditions. Show me how this is likely to happen, and I'll change my mind.

In fact, there is a way in which CO_2 really can kill people, and I've seen some of its horrible effects firsthand. But it has nothing to do with global warming.

In 1985, I was a postdoctoral member of a sediment-coring expedition to Cameroon, West Africa, with my former graduate adviser at Duke

University, Dan Livingstone. With the help of aquatic ecologist George Kling, who was then a graduate student of Dan's, we collected long cores from the thickly jungled crater of Lake Barombi Mbo, a mile-wide bowl of remarkably clear water an hour's drive inland from the sea, which were later used to develop pollen records of local rain forest history. After the work there was finished, George and I headed into the cooler, grassier highlands of central Cameroon to study other more remote sites. One of those was a lovely crater lake called Nyos.

Lake Nyos lay far up a grass-carpeted slope from the deeply rutted dirt road that rings the central highlands, too far for us to lug our heavy sampling gear with ease. This would be more of a scouting trip than a sampling mission. A kindly fellow named Mr. Joseph met us in Nyos village on the lush, fertile valley floor below the lake, and he assigned one of his sons to guide us along the unmarked trail. It was a hot, sunny day, and the lake sparkled at our feet as we rested on the shore and looked out across the smooth blue surface to steep gray cliffs on the far side, about half a mile away. To us, it was a beautiful setting for a relaxing interlude in our sampling schedule.

But if we had been able to bring enough equipment with us to perform the usual analyses from our inflatable raft, we would have discovered several surprising things about Lake Nyos. Though we didn't know it at the time, it is Cameroon's deepest lake; more than 650 feet (ca. 200 m) at the center. And that ample reservoir was supercharged with a natural store of dissolved CO_2 that entered in the form of geologically carbonated groundwater. A little more than a year afterward, on August 21, 1986, a billion cubic yards (a cubic kilometer) of that gas burst forth and killed 1,700 people in the valley below. Among the victims were Mr. Joseph and his family.

As I later described in the September 1987 issue of *National Geographic,* the timing couldn't have been worse. It was the close of market day in Nyos village, and hundreds of people had come in from surrounding villages to buy and sell housewares and garden produce. Night had just fallen, and many had gathered indoors for dinner and sleep.

Hadari, a herdsman who had watched in stunned silence from a nearby hillside as the gas erupted from the lake that night, later described it to me. "It rose up like a white cloud, and then it sank into the valley like a flood," he said. Carbon dioxide is heavier than air, so it poured down on the low-lying villages like water, drowning them in a river of fumes 160 feet (50 m) deep and up to 10 miles (ca. 16 km) long. People in Nyos village suddenly fell away from their crowded dinner tables, struggled for breath, and died on their floors. Others died in their beds or on their doorsteps in the stifling darkness.

The carcasses of 5,000 cattle littered the grassy slopes, and many were still there when I helicoptered over them with photographer Anthony Suau about a month later. Even the vultures and flies were dead, "even small ants," Hadari said. But the grasses and trees were untouched, still as luxuriantly green as ever. To plants, CO_2 is the life-giving essence of air: they breathe it and thrive. To animals, ourselves included, it's a waste product to be exhaled, and it's toxic in high doses. Biomedical labs sometimes euthanize study mice by placing them in a closed container with dry ice in hopes of sending their subjects gently and humanely into permanent sleep, but it's actually not a very nice way to go. High concentrations of CO_2 induce sensory hallucinations, including painful burning sensations that were reported by Nyos survivors, then convulsions and death.

I will always regret having taken the easy route to Lake Nyos on that lovely day in 1985; if we had gone to the trouble of studying the lake properly, we could have warned people of the danger. Instead, survivors later found our names inscribed in Mr. Joseph's guest book and assumed that George and I, being the only foreign scientists in the area before the eruption, had planted a bomb in the lake. We might as well have, for all the good we did for the residents of Nyos village.

And I will always think back on that awful event to remind myself of what a *real* CO_2 catastrophe is like. Next to that, the greenhouse effect just doesn't compare. If we end up emitting 5,000 Gtons of carbon by burning through our coal reserves, then we'll likely trigger an intense and long-lived warm period that would be as severe as a realistic greenhouse future gets. And we should do our best to prevent it, not because it will exterminate us but because our descendants and other living things will actually have to endure it if we let it happen.

How might modern ecosystems and species respond to such a planetary fever? Nobody knows for certain, but we can make some reasonable guesses, and in the next two chapters we'll look to the distant past for glimpses of relatively moderate to supergreenhouse conditions that are revealed in geological records. Hopefully, our grandchildren's grandchildren will never have to face a super-greenhouse anyway. When I last spoke with Archer, he treated it like a thoroughly preventable problem. "We simply can't let it happen," he said. "My personal guess is that we'll probably end up releasing about 1,600 gigatons. In that case, we'll stabilize carbon dioxide levels close to 600 parts per million and, hopefully, avoid some of the very worst impacts."

But whichever path we choose to take into the Anthropocene future, it's now clear that we have already locked ourselves and our world into some uncomfortably large changes. If you're a hard-core fatalist, you might use such points to argue for giving up and doing nothing. That, however, would be a mistake.

Paleobotanist Steve Jackson recently wrote in an editorial for *Frontiers in Ecology and the Environment*, "Climate change may sometimes be inevitable, but that is not a good reason to invite or accelerate it." What we've already done to the global climate system will have surprisingly far-reaching consequences, but those effects are much less extreme and long-lasting than what could happen if we don't reduce our carbon consumption as much and as rapidly as possible. The path to a grim 5,000-Gton scenario lies before us, but we're not inescapably committed to following it just yet.

Our very existence at this pivotal moment in history gives us the amazing ability—some might say the honor—to set the world's thermostat for hundreds of thousands of years, and we have already bequeathed a complex climatic legacy to a long line of future generations. Whether we like it or not, our behavior during this century will determine the magnitude and longevity of that inheritance.

3

The Last Great Thaw

*Is there anything whereof it may be said, "See,
this is new?" It hath been already of old time,
        which was before us.* —*Ecclesiastes* 1:10.

When we look beyond 2100 AD to a deeper future, what we see there
will depend on what happens on this side of that boundary line. If we
limit our carbon emissions to 1,000 Gtons, then we'll avoid the more
extreme environmental changes that could accompany a 5,000-Gton
scenario, but our descendants will still breathe air that's richer in
CO_2 than any inhaled in the history of our species, and global average temperatures may still climb 3 to 5°F (2 to 3°C) on the way up to
thermal maximum.

But how realistic are these scenarios? Has anything like the warming from a 1,000-Gton release ever happened before? And if so, how did
landscapes and living things respond?

Computer models show us what can happen to climates in theoretical simulations, but geohistorical studies show what actually has happened in the past. Both approaches, the abstract and the concrete, have
their strengths and limitations. Models can be constructed to simulate an
infinite range of possibilities, but they may or may not be realistic. Paleo-
ecological records are grounded in reality, but they can't necessarily be
manipulated to answer specific questions; whatever information they
happen to offer is all you get. Both are most powerful when used in tan-
dem, with history guiding the models. After seeing what CLIMBER and
other models do to a virtual Earth when carbon emissions rise by "moderate" amounts, it's also interesting to look for an example of a relatively

moderate warming event that occurred in the past. Together they can sketch a compelling summary of how a version of our current warming trend has played out in a real-world setting and what it can and cannot tell us about the future.

Large-scale warmings have come and gone many times before this, so we have a long list of possible tales from which to choose. However, the biological backdrops of those events stray more and more widely from those of today the deeper we look into the past. Many of them happened before the continents reached their current positions and before most species looked as they do now. But as we probe the deep future, we want to know what living with 550 to 600 ppm of CO_2 might be like, and we want it in the context of a physical world that closely resembles the one we know today.

That constraint leaves us with a much shortened list of possible paleo examples to choose from. If we're going to limit our search to times when most of today's plants and animals shared the planet with us, then we can't look more than a few hundred thousand years back through evolutionary time. Fortunately, more information exists for younger events than for older ones because erosion and other processes have had less time in which to erase the traces they've left behind. Although nothing within the last million years perfectly matches today's 387 ppm CO_2 context, much less the 550 to 600 ppm condition that we're considering in our moderate scenario, there is still much of value to learn from previous warmings. And this choice of a geologically recent time frame also allows us to take advantage of ice cores, the frozen archives of ancient air samples that lie buried in glaciers and continental ice sheets.

The longest such records come from the two largest ice realms on the planet: Greenland and Antarctica. These records are unique in the extreme length and high quality of their layered sequences, and the great thicknesses of those deposits attest to their age; many thousands of annual snowfalls are compressed into those icy stacks. The strata that lie beneath the highest gently rounded domes are our primary targets of study, not just because they are so numerous but also because they move little in comparison to marginal ice, which flows and churns like a slow-motion river under the pull of gravity. Such stability reduces the risk of errors from physical distortions.

At the time of this writing, the longest of all ice core records was produced by the European Project for Ice Coring in Antarctica (EPICA), a multinational effort involving scores of investigators. The core was drilled through 2 vertical miles (3 km) of ice on East Antarctica's Dome C and it covers 800,000 years, long enough to encompass eight cold-glacial, warm-interglacial cycles. The most recent of those cycles are also registered in a 420,000-year-long record from the nearby Vostok station as well as in cores from central Greenland. This similarity of signals among disparate locales lends support to the climatic tales they tell.

Micropockets of ancient air trapped in the cores show that CO_2 concentrations were lowest during cool ice ages, generally hovering near 190 ppm; methane, a rarer but more powerful greenhouse gas, averaged 0.4 ppm then. The highest of the warm spikes in the record barely scraped 300 ppm CO_2, and its methane levels maxed out between 0.7 and 0.8 ppm. To put our modern times into historical perspective, remember that CO_2 concentrations are now 387 ppm and rising, and methane is currently at 1.8 ppm. And even our moderate-emissions scenario is expected to raise those concentrations far beyond anything that the longest, oldest ice core records can tell us about.

Most of the warmings between ice ages also differed from that of today because their primary trigger wasn't the greenhouse effect but changes in seasonal solar heating in the high northern latitudes. Greenhouse gas concentrations did rise during past interglacials, but it was mostly because higher temperatures sped up microbial decay rates and reduced the solubility of gases in the oceans. In our case, we've flipped the arrows of cause and effect by boosting the gas concentrations first. If enough ice remains in the distant future for scientists to read the layers set down during our times, they'll see an unusual break in the long-standing relationships between global temperature and atmospheric composition. In the Anthropocene zones of their ice cores, rising greenhouse gas concentrations will coincide with or slightly precede, rather than follow, the warming trend.

To find the best examples of how life as we know it responded to a major warming in the past, we should turn our attention to events that lie as close to us in geologic time as possible. As luck would have it, a recent interglacial fits the bill nicely. To reach it, we need travel only 117,000 years back in time. *Only* 117,000 years? Nobody but a geologist or paleo-ecologist would put an "only" in front of that number. For those who are used to dealing with vast chunks of geologic time, it's easy to forget that 117,000 years is a very long stretch from the point of view of living things.

Just to make that point more clearly, let's imagine that we're traveling along a path that leads us back through climatic history. We'll use the EPICA ice core's bumpy temperature curve as our trail to the Eemian interglacial.

We feel only minor bumps and jostles in the early stages of that ride. We rumble gently through the wars, civilizations, and inventions of the last millennium. We pass the births of Mohammed and Jesus Christ further along the time track. After we've traversed another 2,000 years, some readers might expect to see Noah's flood, followed by the creation of the world another 2,000 years down the line. But the track continues onward from the 6,000-year mark, rising slightly over four more millennia to the beginning of the warm early Holocene, back during the early days of grain agriculture in the Middle East. Another thousand years pass, and another.

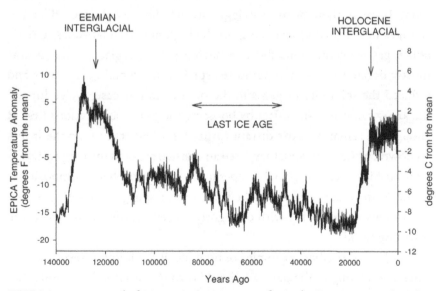

EPICA ice core record of Antarctic temperatures from the Eemian interglacial to today. *After EPICA community members, 2004*

By now EPICA has led us through twelve millennia of cultural and climatic changes. But to reach our final destination, we must also slide down into a cold pit full of sharp-looking temperature spikes that takes nine times as long to cross as the stretch that we've already passed through. We're getting ready to experience, in reverse, a full-blown ice age.

It's a rough ride through the thermal chop of the last glaciation, which scraped much of Canada's rock-and-soil skin off. More than a dozen saw-toothed warmings punctuated it, mostly in response to sudden changes in ice extent or ocean circulation, though none of them lasted more than a few centuries or millennia. As we rattle up the slope of each warming spike only to drop headlong down the older, steeper face, we might long for the more subtle ups and downs of Holocene climate.

Relief finally comes when, after covering 117,000 years of vigorous climate change, we reach the far side of the glacial temperature pit and climb to the edge of the Eemian interglacial. Once again, the path lies high and warm before us for another 8,000 years or so. At this point on our backward journey, most of the Eemian seems to be about as warm as the Holocene was, but there's one last thermal crest ahead; those earliest five millennia were warmer still. Beyond that peak, just past the 130,000-year mark, yawns the toothy canyon of another, earlier ice age. Let's not go there just yet.

To put this into more mundane terms and in normal temporal order, the Eemian interglacial began about 130,000 years ago. During the 5,000 years between full glacial conditions and the early Eemian peak, temperatures in East Antarctica rose by 22 to 25 degrees Fahrenheit (12 to 14°C). This averages out to a rate of roughly half a degree per century, less than the 1.3 degrees (0.7°C) of warming that Earth as a whole experienced during the twentieth century. However, the high, frigid mass of East Antarctica where the EPICA coring site lies is also warming more slowly than the global average today, so it wouldn't be surprising if it lagged behind the rest of the world during the Eemian, too. But despite the relatively slow onset, the EPICA site did eventually became several degrees warmer than it is today for several millennia, and then it settled closer to modern conditions before dropping off into the last great glaciation, about 117,000 years ago.

The warm period's name stems from the Eem River in the Netherlands about 20 miles (30 km) east of Amsterdam, which flows through clay- and sand-rich marine deposits laden with the shells of subtropical mollusks. Today many of those species, such as sharp-pointed needle whelk snails (*Bittium reticulatum*), generally avoid the chilly North Sea and instead favor warmer coasts such as those of the Mediterranean region. I remember collecting hundreds of *Bittium* shells from Turkish beaches as a kid when my father spent a year teaching math in an Istanbul high school. Since their discovery, during the late nineteenth century, the formerly submerged shelly sediments beside and beneath the Dutch riverside town of Amersfoort have represented the classic geologic reference point of the Eemian interglacial.

Technically speaking, the Eemian moniker refers only to European history; in North America, it's been called the Sangamon, and in Russia it's the Kazantsevo. At sea, it produced one of the largest of many oscillations in long geochemical records derived from sediment cores, and so marine geoscientists refer to it as Marine Isotope Substage 5e. But for our purposes, Eemian will do.

Ice core evidence suggests that Eemian CO_2 concentrations stayed fairly close to 300ppm. Presumably, because the greenhouse gas concentrations were lower back then than they are now, global average temperatures should also have been lower. But that's not what the data tell us. Many records put Eemian temperatures higher than ours, typically ranging from 2 to 5°F (1 to 3°C) above today's global average. What's up with that?

Could the gas bubble records from ice cores be wrong? Probably not by much, because the same results appear in core after core. And other historical tools yield similar results, as well. For example, the Scandinavian research team of Mats Rundgren and Ole Bennike has coaxed temperature data from ancient willow leaves that they harvested from Eemian deposits at various European sites. Because plants breathe CO_2, their leaves sometimes respond visibly to changes in the abundance of that life-sustaining gas. The telltale feature in this case was the number of microscopic breathing holes, or stomates, in a standard area of leaf surface; many plants react to rising CO_2 concentrations by producing fewer stomates. From their stomatal index data, these investigators concluded that Eemian CO_2 concentrations hovered between 250 and 280ppm, which is remarkably similar to the ice core numbers.

Could the temperature estimates be wrong? We have several ways of getting at those values, too, most having something to do with stable isotopes (an isotope is a variant of an atom, as a breed is a variant of a domestic dog or cat). One of the most commonly used isotopes in this kind of work is deuterium, a heavy version of hydrogen that contaminates the water molecules in glacial ice. It serves as a paleothermometer because its abundance relative to normal hydrogen changes with temperature. The deuterium records of the EPICA and Vostok ice cores suggest that Antarctic temperatures were about 3 to 5°F (2 to 3°C) warmer than today for the first 4,000 to 5,000 years of the Eemian. But for the rest of the interglacial, temperatures more closely resembled those of modern times.

Ice core records, however, represent only single reference points on a map. To calculate global average temperatures today, we rely upon hundreds of widely dispersed weather stations, but detailed Eemian temperature reconstructions number only in the dozens. So it is misleading to simply say that EPICA data show that the world was X degrees warmer than today during the early Eemian. EPICA's ice layers can't tell you exactly what the global average was back then, just as a modern weather station situated at the coring site can only register the local conditions in that part of Antarctica. And the ice records only preserve information during seasons in which snow accumulated; if little or no snow fell during certain months of the year, then much of the annual record might not be represented at all.

The Mount Moulton horizontal ice core trench, which for now represents our only window on the Eemian climates of western Antarctica, displays just such geographic variability. It lies across the Ross Ice Shelf from the much larger eastern sheet where the EPICA and Vostok cores were drilled. Deep ice can ooze sideways under its own weight, and in the vicinity of Mount Moulton it presses forcefully against the base of the mountain and curls itself upward like layered taffy. If you walk outward from the rock-ice contact zone, you can traverse half a million years of polar history in parallel strips of translucent blue strata. Rather than drilling down through that archive, investigators from several American universities have simply sawed long sampling trenches through it.

Today, Antarctica is warming far more dramatically in the west than in the east, and that kind of regional variability also typified the

Eemian. Although the Mount Moulton area warmed about as much as East Antarctica did 130,000 years ago, the subsequent cool-off there was more gradual than it was elsewhere, more of a smooth slide than a cliff dive. The basic storyline was the same: an abrupt temperature rise, several millennia of warm conditions, and a return to glacial cold. But the finer details vary from site to site, reminding us to be careful when using single records to study global history.

Our present warming trend is more intense in most of the polar regions than it is at lower latitudes, and the same situation existed in the past, too. Much of the Arctic was 7 to 9°F (4 to 5°C) warmer during Eemian summers than it is today, while the global average, as best we can tell, was slightly less than half of that. So if we rely only upon high-latitude records for estimates of past global temperatures, then we may bias our guesswork toward the high end of the scale.

But even if the Eemian Earth was only slightly warmer than today, greenhouse gas concentrations were definitely much lower. Does this mean that we're wrong in linking modern global warming to carbon emissions? Not at all. We've simply gotten things backward nowadays. Today, CO_2 and methane are driving temperatures up, but back then it was warming itself that drove the gas buildups. As the orbital cycles ran their appointed course, the loss of reflective snow and ice from northern high latitudes let darker lands and seas soak up even more incoming solar energy. That, in turn, would have accelerated microbial respiration rates and increased the activity of methane-belching wetlands, wrapping the planet in an even thicker insulating robe of heat-trapping gases. In our case, the insolation cycles that produced the Holocene interglacial of the last 11,700 years have already passed their peaks, and the present warming is mostly of our own making.

Despite its different origin, the Eemian example clearly shows that even moderate heating can demolish plenty of ice; indirect but compelling signs of interglacial melting appear in stranded fossil deposits all over the world. Among the hundreds of exposed paleo reef sites that formed during the Eemian, the most reliable indicators of past sea level are those found on coasts that are geologically stable and therefore unlikely to have lifted or lowered the ancient deposits. One such site has been described in a study led by Paul Hearty, a geoscientist from the University of Wollongong in New South Wales, Australia. Hearty and his colleagues examined a wave-cut bluff on Rottnest Island, a low-lying chunk of arid land rimmed with white sand beaches just offshore from Perth, and found fossilized corals of early Eemian age sitting at or above head height beyond the surf zone. Because such corals prefer to live in shallow water, this discovery demonstrated that sea levels of the time stood 6 to 10 feet (2 to 3 m) higher than they do today.

The longer the warm spell persisted, the higher sea levels continued to climb, even after the early thermal peak had passed. Moving farther east onto the mainland, thereby retracing the encroachment of the swelling interglacial ocean, Hearty's group found petrified mollusks still attached to their parent rocks and to each other 23 feet (7 m) or more above modern sea level. These were creatures that lived and died several thousand years after the Rottnest corals did, and their higher, more easterly positions reflected a long-term sea-level advance. Together with similar findings elsewhere in the tropics, these results help to sketch a coherent sequence of Eemian sea-level change.

At the start of the interglacial, ocean surfaces rose 6 to 10 feet (2 to 3 m) higher than today and stayed there for several thousand years. During the second half of the warm phase, sea level continued to rise in several steps to a maximum of 23 feet (7 m) or so, probably in response to partial collapses in lingering polar ice sheets. A long-term decline then followed as the next ice age withdrew water from the oceans and froze it back onto land.

Then, as now, the largest ice deposits lay in the Arctic and Antarctica, and these are the most likely sources of the mid-Eemian meltwater pulses. Something akin to the volume of Greenland's ice sheet could do the job, but we can safely say that Greenland wasn't the only donor to the water budget. Local ice cores there show that most of the central dome survived those long millennia of warmth, leaving a sizable remnant that was at least half as extensive as today's pile. On the opposite end of the planet, the main mass of the East Antarctic ice sheet also survived; otherwise, we wouldn't have those superlong records from EPICA and Vostok.

This means that the interglacial sea-level rise was probably the work of several ice sources, most likely a joint effort between Greenland

Fossil oysters near Durban, South Africa, still attached to rocks that were once submerged during the Eemian interglacial, now stranded well above sea level.
Curt Stager

and West Antarctica. It also means that even 13,000 years of heating didn't destroy the Greenland ice sheet altogether, which might offer some reassurance as we watch temperatures and melting rates rise there during this century. If we can limit future warming to something akin to the Eemian situation, as would be expected for a moderate-emissions scenario, then perhaps we can also keep a fair bit of our terrestrial polar ice in place.

On the other hand, sediment cores collected from offshore sites north of Greenland show that most of the Arctic Ocean was seasonally ice-free then. Planktonic microorganisms of the sort that live under sea ice today gave way to other forms that are now common in ice-free waters farther south, and stable isotopes of oxygen preserved in the empty shells of those long-buried creatures also tell of open water near the pole. But the loss of that floating ice cap wouldn't have changed global sea levels—it's only land-based ice that does so.

Since there was nobody to chart coastlines precisely back then, we don't have a complete picture of how sea-level rise reshaped maps of the Eemian world, but we do have some general insights into what was lost. Saltwater covered much of northern Europe and the west Siberian plains, isolating Sweden and Norway from the mainland as a sausage-shaped Fennoscandian island. And it's a safe bet that many other low places, including those icons of modern sea-level rise, coastal Bangladesh and some of the smaller Pacific islands, went under for at least part of the Eemian.

Apart from warming and sea-level rise, some geological records also show what the Eemian did to rainfall. Moving farther east from Rottnest Island to what is now the hot, dusty heart of central Australia, we find an Eemian weather surprise. Increasing summer warmth invigorated monsoons throughout most of the lower latitudes, and this brought more rainfall to the arid outback. In the Kimberly region, the Gregory Lakes overflowed their banks, and the crusty flats surrounding what is now ephemeral Lake Eyre—often more of a desert mirage than a lake—were flooded continuously. Across the Indian Ocean in the highlands of East Africa, heavy rains overfed the Nile and poured river runoff into the eastern Mediterranean, where it capped the briny sea with a layer of less saline water. That buoyant lid prevented formerly dense, salty surface water from sinking or stirring deeply under the influence of the winds, and it therefore cut off the supply of dissolved oxygen to the bottom. This produced thick layers of slimy organic ooze that are still preserved in marine sediment cores collected near the coast of Egypt. And copious rains over the Sahara helped to imprison what are now vast seas of drifting sand under an emerald carpet of grass and trees.

The warmer and generally wetter interglacial world was inhabited by many animals that are familiar to us today; most of its dominant plant species are still with us as well. Although some notable evolutionary transformations have occurred since then, most of the largest biotic differences between now and then reflect environmental factors rather than genetic mutations. Mammoths are absent from our landscapes because of human hunting or the human-driven destruction of habitats by fire, not because they've turned into a less hairy form of elephant. The time periods that separate us from the Eemian are too short even to have fully fossilized many of the bones, shells, and foliage left behind by its residents, which can seem more like mummies than stone replicas when discovered. Because of this similarity to modern organisms, we can use the remains of Eemian biota, not only to infer the effects of past climates on living things, but also to learn about the climatic conditions themselves.

During the warmest early phase of the Eemian, boreal treelines lay hundreds of miles north of their present locations, often pressing right up against the coastlines of the polar sea. Much of Baffin Island and southern Greenland were cloaked in white birch woods, and hazels and alders rustled in the breezes of northern Sweden and Finland well above the Arctic Circle. Pollen grains in lake and peat deposits show that great forests of oak, hornbeam, and yew covered much of Europe north of the Alps. Farther east, the pollen of spruce, fir, and pines blew into the crystal clear, mile-deep waters of Lake Baikal during most of Siberia's Eemian (or Kazantsevo, if you prefer), which means that local boreal taiga woodlands of the time looked a lot like they do now. However, to the north of Baikal, evergreen forests eliminated coastal tundra while broad-leafed trees such as oaks and elms invaded the southern borders of the taiga belt.

In North America, too, the main theme was poleward migration. In central Alaska and the Yukon Territory, spruce-birch woods much like those of today moved in over retreating tundra as local weather became warmer and wetter in summer, eventually reaching much farther north than the Arctic treeline extends now. Cedar-hemlock-fir forests claimed the Pacific coasts of Washington and British Columbia, and the central plains sprouted a complex mosaic of deciduous forest and savanna. In Florida and Georgia, dry oak woodlands expanded during the opening phase of the Eemian before giving way to pinelands and cypress swamps. And in upstate New York and southern Ontario, trees that are now more typical of the southern Appalachians kept the local bears and squirrels fat with generous supplies of hickory nuts and acorns.

The Asian summer monsoons, which had been suppressed by glacial cold for thousands of years, reawakened with the return of interglacial warmth, and heavy weathering by seasonal heat and copious rains produced thick layers of soil on the formerly dust-covered plateaus of central China. Pollen grains collected from lake deposits in the rugged mountains of Colombia show that thick, humid forests of oak and glossy-leafed, flow-ering *Weinmannia* returned to the high country after temperatures there climbed 1 to 2 degrees higher than those of today.

When the cold began to creep back in about 117,000 years ago, grassy steppes pushed the taiga back from Baikal's shores, and leafy de-ciduous forests retreated southward as conifers and tundra reinvaded the increasingly icy landscapes of Europe. Northern Germany was buffeted by dust storms and brush fires as the climate became cooler and drier. The vegetation of central France began to resemble that of northern Scandinavia as gigantic ice masses re-formed at high latitudes and alti-tudes.

In general, changes in animal life during the Eemian paralleled those among the plants; herbivores followed their food supplies and pred-ators followed their prey as warming reshaped the home ranges of spe-cies around the world. A faunal checklist for the Eemian forests of central Europe includes wild boars, wolves, foxes, hares, beavers, martens, and a rich assortment of mice. The descendants of tiny water voles that had scampered and paddled about northwestern Europe during the previous interglacial also reinvaded the territories of their forebears as their more cold-tolerant cousins headed poleward.

It can sometimes be more difficult to reconstruct the ecology of Eemian animals than that of the vegetation because you need to find their bones or cuticles or shells in order to learn much about them. When those kinds of remains are available, though, they can be very informa-tive. We know, for example, that New Zealand was 3 to 5°F (2 to 3°C) warmer than now because diagnostic species of beetles were well pre-served in lake sediments from that time period, according to the work of paleoecologist Maureen Marra at the Victoria University of Wellington. Plants, on the other hand, are more commonly studied because their pol-len grains are widely dispersed by wind and water; pollen and other plant remains in New Zealand lake sediments confirm the beetle evidence for warming and also show that the weather there was rainier than it is now.

In the case of animals, you also might be fooled into thinking that a certain kind of creature evolved or went extinct simply because it ap-peared or disappeared at a given location. But we now have enough study sites to show that most Eemian animals simply followed their preferred vegetational settings as climates changed, vanishing in one place and show-ing up again in another. And some of the migration-related changes were nothing short of spectacular.

At the height of the warming, hippopotami splashed and snorted in the Thames River not far from the future city limits of London. Rhi-nos stomped the British underbrush, straight-tusked *Elephas antiquus* munched foliage as far north as Denmark, and water buffalo dipped their heavy crescent horns to drink from the Rhine. Although most of Europe was warmer during the Eemian than it is now, particularly in summer, it wasn't all that much warmer. Perhaps hippos, rhinos, elephants, and water buffalo could also thrive in modern Europe if we gave them a chance to, especially now that we're back on a long-term warming trajectory.

In North America, the largest lions of all time—cave-dwelling *Pan-thera leo atrox*—shared verdant landscapes with even larger short-faced bears. Enormous beavers, *Castoroides ohioensis*, gnawed riverside trees in the Midwest as mammoths, wild native horses, and several species of bi-son looked on. Mammoths and their kind are sometimes called "ice age mammals," but they actually lived in much of North America and Eurasia during interglacials, too. Such cold-tolerant beasts merely retreated into shrunken remnants of their favored habitats when it warmed, and many of them lived through the Eemian successfully enough to become familiar to us thousands of years later—polar bears, various ice seals, and Arctic foxes among them. To record most of the major changes in species over the last few hundred thousand years at a single fossil site is to watch cli-matic shifts of the past drive animal and plant communities in and out of the area, with more dramatic effect than the push of genetic evolution.

This raises an important point when we compare the Eemian ex-ample to modern times. Although few of us realize it, we live in a species-depleted Anthropocene world, and the full tallies of animals that are now artificially missing could fill many pages in this book. The Americas have lost their sabertooths, cave lions, giant ground sloths, rhinos, and multiple species of elephants, bison, and camels. Australia no longer has its short-faced kangaroos or giant wombats, and New Zealand no longer has moas. No heavy-racked Irish elk live in Ireland today, and no cave bears haunt the painted caves of France.

These absences aren't simply the legacy of heating at the end of the last ice age; most such species survived multiple warmings in the past, and not just interglacials but also the many abrupt, century-scale hot-cold shifts that disrupted otherwise icy millennia between the Eemian and the Holocene. As all but a small minority of scientists now agree, those worldwide disappearances were more the result of human activity than of climatic change.

Sadly, from the perspective of many of today's remaining species, it's only going to get worse from here on out. Even if we take a relatively moderate emissions path into the future and thereby hope to avoid destroying the last polar and alpine refuges, warming on the scale of an Eemian interglacial will still nudge many species toward higher latitudes and elevations. In the past, species could simply move in response to the push and tug of climatic changes, but this time they'll be trapped within the confines of habitats that are mostly immobilized by our presence. People, more than climate alone, will determine their fates as the Anthropocene continues to unfold.

And what about the early ancestors of those people? What were they doing during that warm respite before the last ice age, back when the plains of North America and the steppes of Asia still supported herds and packs and prides so rich in abundance and diversity as to rival the Serengeti? We have only scattered snapshots of Eemian humanity to go on, for several reasons. For one thing, no humans lived in the Americas or Australia back then because they hadn't crossed over from Asia yet. For another, Eemian-age deposits are unevenly distributed over the planet; they can be difficult to identify; and they must be conveniently exposed to study, usually by well-placed quarries or roadcuts.

But most importantly, there weren't many humans anywhere on Earth back then, at least not outside Africa. The Eemian warm period, whether by chance or by causation, marked one of the earliest known dispersals of modern *Homo sapiens* out of the African homeland.

That's not to say that early hominids didn't already occupy much of the Old World then. Some diminutive, hobbitlike beings wandered the moist, rain-forested islands of Indonesia. In what is now France and Germany, spear-toting Neanderthals hunted woolly tuskers and reindeer on "mammoth steppes" under glacial conditions. Later, Eemian Neanderthals switched to elephants, rhinos, brown bears, deer, and steerlike aurochs in dense woodlands as open-ground prey shuffled north with the retreating cold, and on coastlines as far south as Gibraltar and the Middle East they harvested mussels, seals, dolphins, and fish. But these were not precisely human in the strictest modern sense despite their tool use and despite new evidence that some of our ancestors shared genetic material with Neanderthals.

Anatomically modern humans, our most direct ancestors, arose in Africa roughly 200,000 years ago and spread to the rest of the world later on. Many scientists believe that the large mammals of Africa have outlasted their relatives on other continents because they coevolved with humans and were therefore less likely to be taken by surprise when two-legged predators approached them with killing tools in hand.

During the Eemian phase of the Middle Stone Age, one of the first waves of fully human wanderers crossed from the Egyptian mainland eastward into what we now call Israel and Jordan. Geochemical records from laminated cave formations show that stronger regional rains of the time opened a corridor to the rest of the world for some of our ancestors by making a trek across the Sinai and Negev deserts more of a hunting trip than a waterless death march. Climate change closed that door, too, at the end of the Eemian when the return of cold climates shut the Middle Eastern rains down. The early emigrants vanished from historical view then, perhaps by returning to Africa or by dying out.

Meanwhile, back in the homeland, the bulk of humanity went about its business as usual. On the edge of the Red Sea in today's Eritrea, teardrop-shaped hand axes litter ancient beach and reef deposits where Eemian humans once feasted on oysters and crabs. On the opposite end of the continent, at South Africa's Blombos Cave, coastal cave dwellers flaked stone and bone artifacts and munched fish and mollusks. Not far away, at Klasies River, people gathered edible plants and shellfish and also hunted penguins—not because it was unusually cold there but simply because native penguins waddled about on the beaches of the South African Cape region then as they do today. When it cooled off again at the end of the Eemian, sea levels fell so low that the caves at the mouth of the Klasies River overlooked a broad coastal plain rather than a food-rich shoreline. They were only rarely occupied after that until about 3,000 years ago, when rising sea levels once again brought the waves back to within easy walking distance from home.

So there we have it, a taste of what happened the last time a world much like our own warmed above modern temperature ranges. The Eemian example does not show us what a super 5,000-Gton greenhouse could be like; in the next chapter we'll look much farther back in time in order to find such an example. But for a glimpse of some basic features of a fairly moderate 1,000-Gton future, it does nicely.

Greenhouse gas concentrations rose in response to the Eemian heat, but they didn't reach nearly as high as they now have with our help. Northern summer warming strengthened tropical monsoons, and sea levels climbed 23 feet (7 m) or so higher than today. Much or all of the Arctic Ocean became open in summer, so we can be fairly certain that continuing our warming trend today will open it again. However, the land-based ice sheets on Greenland and western Antarctica still kept much of their bulk in frozen form, and the eastern Antarctic ice sheet shrank even less.

Eemian warmth sped Arctic warming up to rates considerably faster than those at lower latitudes, almost certainly as a result of amplification processes such as changes in surface reflectivity and the decay of organic matter—changes that are also operating today. However, not all permafrost succumbed to the prolonged thaw, probably because surface materials insulated it; some of the subterranean ice in Alaska is 750,000 years old, according to a study by Canadian geoscientist Duane Froese and colleagues, which means that it has survived multiple interglacials. The Eemian warming did push tundras, steppes, and forests poleward along with their resident creatures, and the subsequent ice age pushed their descendants back again in an early version of the climate whiplash that now lies ahead of us today. In a wide-open world, it's not much to ask of a mobile organism to shift its geographical range over the course of generations as long as the necessary food and habitat conditions move with it, and many species did this over and over again without obvious difficulty during previous cold-warm-cold oscillations.

But much is different nowadays. Artificially generated greenhouse gases are driving the changes now, and because they operate beyond the limits of any particular latitude, season, or time of day they may destroy more ice in less time than the preceding interglacials did. In addition, the role that our ancestors played in the Eemian world was relatively simple and markedly different from the one that we play in this one. Even though they hunted and gathered widely, those few and technology-poor early peoples generally had less impact on their surroundings than the average beaver colony or mammoth herd—at least until Clovis-age cultures brought heavy spears and other deadly factors into play at the end of the last ice age. But today we spread settlements, farms, factories, and roads far and wide. Our complex technologies allow us to reshape landscapes and relocate or exterminate entire species. Nowadays in this remarkable Age of Humans, we are as much a barrier to biological migrations as high mountains, swift rivers, and broad oceans have been in the past. As Anthropocene warming rises toward its as yet unspecified peak, our long-suffering biotic neighbors face a situation that they have never encountered before in the long, dramatic history of ice ages and interglacials.

They can't move because we're standing in their way.

4

Life in a Super-Greenhouse

*The farther backward you can look, the farther
forward you are likely to see.*

—Winston Churchill

We've now glimpsed what a recent warm period was like, using the
Eemian interglacial as a rough guide to what a relatively modest 1,000-
Gton carbon release could bring us. But none of the events that are
recorded in ice closely resemble the super-greenhouse that we could
unleash by burning most of our remaining fossil fuel reserves. Just how
extreme can Anthropocene climate change become, and what might life
in such a world be like?

The details of what it would take to drive climate over a thermal
cliff are still unresolved, but history makes one thing clear: the critical
tipping point certainly exists because it's been passed before. Not during
the last 130,000 years, though, and not during the lifetime of most species
alive today. It happened 55 million years ago, some 10 million years after
the demise of the dinosaurs, and it currently represents one of our best
historical examples of what radical Anthropocene warming could be like.
Unlike the Eemian, it was not caused by orbital cycles, and the mechanisms driving it were global rather than hemispheric in extent. But above
all, the case that we're about to consider here was certainly extreme. By
all accounts, it was one of the most abrupt and intense greenhouse warmings ever to occur on this planet.

In order to understand how such a superhothouse could happen
long before humans entered the picture, we'll have to look back into the

earliest phase of the Cenozoic era, which began 65 million years ago and continues today through the subset of geologic time that we are now calling the Anthropocene epoch. That great era is sometimes described as the Age of Mammals in acknowledgement of the tremendous diversification of hairy, warm-blooded, milk-producing critters that defines it.

The first 31 million years of the Cenozoic, dubbed the Paleocene and Eocene epochs, were already much warmer than today for reasons that are as yet not fully understood. Some geoscientists attribute it to different patterns of ocean current flow around a slightly different orientation of drifting continents. For example, the Central American land corridor didn't exist yet, so tropical marine currents and their associated climatic effects were probably somewhat different. Most experts, however, link the warmth to greenhouse gas concentrations that were much higher than they are today—though, again, the causes remain unclear.

For whatever reason, something kept nudging Earth's temperatures higher and higher during the early Cenozoic. During the Eocene climatic optimum, beginning around 50 million years ago, global average temperatures were 18 to 22°F (10 to 12°C) or more above today's mean for several million years. From that point on, the prevailing direction of climate change has mainly been toward inexorable cooling. Roughly 34 million years ago, the first permanent ice sheets began to form in Antarctica. The Arctic followed suit about 8 million years ago, and during the last 2 to 3 million years the planet has endured dozens of ice ages. What we face now as we contemplate the onset of an extreme hothouse in the Anthropocene future is, in essence, a sudden reversal of geologic history. If we take the 5,000-Gton emissions path, then we'll essentially jerk global climate back into the early Eocene.

This outline of events is oversimplified, though. The patterns of long-term change were far from smooth, particularly during the early Cenozoic, when brief surges in the warming trend drove several hot spikes into the rising temperature curve. One of the most notable and dramatic of these occurred 55 million years ago, and it is considered by many geoscientists to be our best real-world example of a worst-case greenhouse future. For something close to 170,000 years, that Paleocene-Eocene thermal maximum, or PETM, forced the world into an exceptionally heated state that bears a striking resemblance to our extreme-emissions scenario.

As of today, we've vaporized about 300 Gtons of fossil carbon. By comparison, most investigators estimate that at least 2,000 Gtons of carbon flooded the atmosphere during the PETM. Some put it as high as 5,000 Gtons, which is close to the magnitude of our worst-case emissions scenario. What could have caused such a thing? For once, we're off the hook; even our earliest humanoid ancestors wouldn't show up in the fossil record for another 50 million years.

To address that issue of origins, we can look for clues in the patterns of warming that the gases produced. And to do that, we'll need information sources other than the ice cores that told us so much about the Eemian interglacial. Our longest ice records reach less than a fiftieth of the way back to the PETM, and the entire early Cenozoic was so hot anyway that it left no ice sheets behind to tell us anything.

Instead of ice, we must turn to sediments for guidance. For example, long cores that scientists pull from ancient marine deposits can contain rich stores of information in the form of tiny protozoans called "foraminifera," or forams. Forams are saltwater amoebas but, unlike the typical biology-lab blobs, they build beautifully coiled or lumpy shells out of calcium carbonate. Thus armored against small predators, they drift or creep throughout the world's oceans like microscopic turtles, and when they eventually die and sink to the bottom their hard, empty shells may remain preserved for millions of years.

Forams are especially useful to paleo-oceanographers because they contain oxygen-18, a heavy, stable isotopic form of oxygen atom. The ratio of that isotope to the more abundant oxygen-16 atoms in a foram's shell reflects the temperature of the waters in which the little creature lived. In that sense, oxygen isotopes can serve as paleothermometers, though their accuracy is blunted when ice sheets selectively trap normal oxygen-16 and hold tons of it away from the oceans. Fortunately for our purposes, the early Cenozoic was largely ice-free, so the oxygen thermometer reads more clearly than it does for samples deposited during recent ice ages; temperature estimates that predate the buildup of ice in Antarctica roughly 34 million years ago are therefore more reliable than the ones that follow it.

Foram data from Pacific cores suggest that tropical sea-surface temperatures jumped as much as 5°F (3°C) above their already-warm

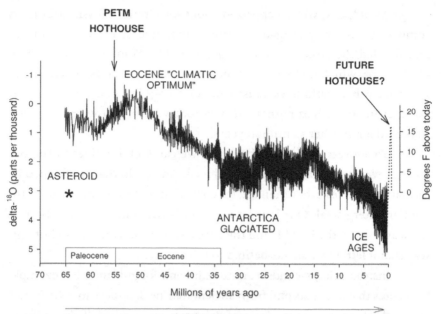

Deep-sea oxygen isotopes and temperatures throughout the Cenozoic Era, showing early warm conditions followed by long-term global cooling. *After Zachos et al., 2008*

state during the opening centuries of the PETM; other temperature indicators in those same cores and from Indian Ocean sediments double that jump. But even more dramatic are the results gleaned from fossil forams of the polar seas. They show that the surface waters of the Arctic and Antarctic warmed by 14 to 18°F (8 to 10°C) within a few thousand years. One estimate puts sea-surface temperatures at the North Pole in the 70s Fahrenheit (23 to 24°C) or even higher. In other words, a time-traveling visitor to the early Eocene could swim in a perpetually ice-free Arctic Ocean with barely a shiver.

Sediment cores lifted from the sea floor near the North Pole by Yale geoscientist Mark Pagani and colleagues also contain substances, called "tetraether lipids," that were once embedded in the oily cellular membranes of floating plankton. Like melted butter after you put it in the fridge, membrane oils can stiffen if the surrounding habitat cools off and they can turn runny if it warms too much. Certain kinds of tetraether lipids help cells to resist those harmful changes by keeping membranes at optimal fluidity as local temperatures vary. By analyzing the lipid contents of polar ocean cores, Pagani's team found that Arctic sea-surface temperatures ranged from 65 to 75 degrees (18 to 24°C) during the PETM, thus independently supporting the foram data.

We don't have a dense, worldwide array of ancient weather stations to tell us exactly what global mean temperatures were during that thermal maximum; as in the case of the Eemian, we have to rely on a relatively small number of reference points. We do know that the tropical seas warmed quite a bit along with the polar regions; according to geochemical data from cores drilled off the coast of Tanzania, temperatures there were as high as 95 to 104°F (35 to 40°C). As best we can tell from widely scattered geological study sites, the PETM heated the planet as a whole by something close to 10°F (5 to 6°C) within several thousand years, and it apparently displayed much the same latitudinal asymmetry that we're experiencing now, with the poles warming more than the tropics. Today, we attribute this to an extra thermal kick as reflective snow and ice give way to darker, heat-absorbing surfaces, but we have no direct evidence of extensive polar ice back then. So why the enhanced polar warming? Perhaps it means that some snow and sea ice still formed in winter. If it kept some locations unusually cool by reflecting sunlight away, then losing those last reflective surfaces when the PETM began might explain the extra jolt of polar warming.

With no gas bubbles from ice cores to guide us that far back in time, climate scientists face major challenges as they try to figure out what produced such warming in a human-free world. But all investigators agree that carbon-based gases, that is, CO_2 and methane, were central players, just as they are today. The evidence for this rings strong and clear in PETM-age sedimentary deposits: a worldwide dip in the abundance of carbon-13, a rare version of carbon that's found in the bodies and waste products of all living things. Only a colossal intrusion of CO_2, methane, or some combination of the two into the atmosphere and oceans could have caused it.

Not only do geoscientists use this distinctive "carbon isotope excursion" to infer the causes of the PETM; they also use it to identify the warm interval itself in rocks and sediments and even in the teeth of extinct animals and the leafy remains of early flowering plants. The carbon-13 dip is a remarkably informative chemical label that allows you to match the ages of samples from one continent to another more precisely than radiometric dating methods would allow, thereby confirming that extreme environmental changes occurred synchronously all over the planet.

Here's how it works. Every living thing is full of carbon, a tiny fraction of which is carbon-13 (also written as ^{13}C). That's because plants inhale ^{13}C-contaminated carbon dioxide from the air around them, although they avoid doing so as much as possible. Because that selectivity is imperfect, plants do contain traces of unwanted ^{13}C, but not nearly as much as you'd find in a comparable mass of molecules drifting about in the open air. The same situation holds true for animals, ourselves included, because grazers eat plants and thereby recycle their carbon through the world's food webs. Because most living bodies are thus well scrubbed of ^{13}C, an unusually low concentration in a geological deposit means that it probably contains the remains of long-dead organisms.

One of the first scientists to recognize the PETM's distinctive ^{13}C signature in a sediment record was Lowell Stott, now a paleo-oceanographer at the University of Southern California. As a graduate student back in the late 1980s, Stott was studying South Atlantic cores and wondering why the ^{13}C values in his ancient forams were coming out so low.

"I got these bizarre numbers that didn't make sense to me," he told me recently. "I thought I might have made a mistake, so I reran the samples, but I still got the same result. Then I tried another species of foram and got the same result again." His advisor, James Kennett, didn't know what to make of it, either. It was both the dread and the dream of any graduate student—a finding that's difficult to explain but too important to ignore. "I was naive enough to be excited by it, though," Stott chuckled. "Nobody had ever seen this kind of thing before." Three years later, Kennett and Stott published the discovery in *Nature*.

In follow-up studies by others, similar drops in ^{13}C values were found in PETM-age deposits from all the major ocean basins as well as on land. And the most reasonable explanation for that global ^{13}C depletion was a massive release of organic fumes, either CO_2 or methane, from decaying peat or other carbon-rich deposits. But which gas was it, where was it hiding before the PETM, and why did it emerge so suddenly?

If CO_2 was the culprit, then it should also have acidified the oceans because it produces carbonic acid when it dissolves in water. Marine sediment cores are well suited to resolve such a question, in addition to preserving the ^{13}C excursion as a convenient time marker. A typical deep-sea core is a cylinder of moist gray to brown mud that can sometimes be so full of chalky carbonate particles that it fizzes on contact with strong vinegar. But the stuff deposited during the PETM forms a strikingly obvious reddish band ranging from a few inches to a foot or more thick because the pale carbonates have been eaten away, leaving behind a rusty claylike residue.

The transition across the base of the red zone is abrupt, indicating a sudden acidification of the deep sea, but the return to normal carbonate deposition was gradual, lasting between 50,000 and 200,000 years. This pattern would be consistent with a dramatic CO_2 rise followed by a long slow drawdown, much like the extreme 5,000-Gton carbon emissions scenario that we're hoping to avoid now. In the context of our current situation, it presents a stern warning; this sort of thing really can happen to our future Earth because, quite clearly, it has already happened before.

One reasonable explanation for the initial CO_2 release stems from seemingly unrelated work on the geological history of the North Atlantic. Great crustal cracks along the midline of that widening ocean basin have been pushing North America and Europe farther and farther apart throughout the Cenozoic era, but during the late Paleocene an unusually active spreading zone opened up between Greenland and what would later become Scandinavia. For hundreds of thousands of years, huge floods of glowing lava burst from the submarine cracks and seared their way into carbon-rich sedimentary deposits. If highly organic or limy materials on the deep-sea floor were burned in this manner, then they would have released CO_2 just as our fossil fuel combustion does, and because they contained the remains of dead marine organisms the resultant vapors would also have been depleted in ^{13}C. If enough such gases were released into the atmosphere and oceans, then they could have caused the greenhouse warming as well as the global decline of ^{13}C concentrations.

But what if the PETM's main pollutant gas were methane instead of CO_2? Methane rapidly oxidizes into CO_2 in the atmosphere, so a methane burst might indirectly cause carbonic acid pollution of the oceans as well. The presumed source? Microbially generated methane ice that accumulates in certain kinds of wet sediments, especially on or near marine continental shelves. Also called "clathrates," these odd substances form when sediment-dwelling bacteria release methane as metabolic waste. Under the right combination of cool temperature and high pressure, bacterial methane can become trapped in tiny cages made of loosely linked water molecules, forming delicate crystal lattices that resemble an unstable version of dry ice. Haul a mud-caked chunk of white methane ice up to the sea surface and put a match to it, and it burns like a candle.

Methane ices are so unstable that any number of triggers, from sea-level changes to climatic warming or volcanism, could presumably have launched a carbon-rich gas assault on the PETM atmosphere. If one or two thousand gigatons of methane emerged over a short time period, ^{13}C concentrations around the world would drop more steeply than if an equal volume of CO_2 escaped. Bacterial methane contains even less ^{13}C than most other biological substances do, so it would take less of it to produce the abrupt ^{13}C excursion than if CO_2 were the only operator.

James Kennett has used this hypothetical mechanism, sometimes called a "clathrate gun," to explain not only the PETM greenhouse but also many of the noteworthy climatic swings of the last hundred million years. The idea is fairly simple. Clathrates build up in oceanic muds, peatlands, and permafrost like a growing arsenal of air rifles loaded with methane charges. Occasionally, the gases burst from their barrels to trigger a global warming until natural processes eventually consume or rebury them.

The clathrate-gun hypothesis has gained favor among those who expect our own greenhouse to cause a similar blowout in the future. But there are problems with it, too. A 1,000 to 2,000-Gton methane pulse could explain the global ^{13}C dilution, but to some investigators the magnitudes of the warming and ^{13}C excursion together seem to be more consistent with a larger, 5,000-Gton slug of CO_2. In addition, David Archer suggests that most of the methane ices in place today are too diffuse and well insulated by thick blankets of sediment to respond suddenly to most environmental changes. If he's right, then the clathrate gun has a safety catch on it. Archer envisions only gradual releases spread over thousands of years, more of a squirt gun dribble than a blast.

Whatever really caused that initial gas spike, it almost certainly set off a cascade of follow-up releases; this could very well happen during a 5,000-Gton emission of our own making, too. This kind of additive, self-amplifying process is referred to by specialists as a positive feedback loop.

Biomechanics expert Steve Vogel, one of my mentors at Duke University, used to describe such feedback loops in a way that caught the attention of all but the most inattentive of his undergraduate nonmajors. "Imagine lying in bed next to the partner of your dreams," he'd say, savoring the sudden lifting of drowsy heads. "You've each got an electric blanket to cover you in the cold, unheated room, and you each have a temperature control knob in your hand. Unfortunately, however, in the darkness and distractions of the moment, each of you has grabbed the other's control knob."

Once the expected chuckles had subsided, he would continue. "Now what happens as the night wears on and the room grows cooler? You feel chilly so you crank the knob up a bit. But that only warms your partner's blanket, and when they feel too warm they turn their own knob down a bit more. That makes you feel even colder, so you turn your knob to an even higher setting, and around and around it goes. Pretty soon, you're dying of cold while your partner is dying of heat. That's what we call a positive feedback loop, not because it's positive in a good way but because its additive effects grow and grow in a self-amplifying cycle."

Applying that memorable concept to the planet, we see that pushing temperatures up too far can unleash positive feedbacks that keep cranking the global thermostat higher and higher. From that point on, the warmer it gets, the more it stimulates processes that release even more heat-trapping gases.

In the case of global warming, there is a long list of things that could become participants in such feedbacks. Warmer water holds less gas in solution than cold water does, so marine heating can squeeze more CO_2 out of the oceans and into the air. A hotter atmosphere also becomes more humid because it sucks moisture from the soils, lakes, and seas beneath it, and water vapor is a major greenhouse gas in its own right. Heating speeds the gas-producing decay of organic matter in wetlands and thawing permafrost, and it can also destabilize frozen methane. By some estimates, the carbon content of today's clathrate methane inventory may approach that of all other fossil fuels combined, so the PETM example makes a strong case against pushing future temperatures any higher than we have to.

And what was the world like in that superhothouse? For one thing, the PETM finished off whatever land-based ice may have persisted near the poles through the already-warm Paleocene epoch. With sea temperatures in the far north in the 70s Fahrenheit, the Arctic Ocean became a tepid, brackish lake. This was a world without ice caps or extensive glaciers; if any snow fell at all, perhaps on the highest circumpolar mountains in the long darkness of midwinter, most or all of it would probably have melted during the warmer, sunnier months of the year.

To tease those insights from ancient marine oozes, paleo-oceanographers have focused less on forams than on the remains of tiny planktonic algae. Unlike forams, those free-floating microbes were more like miniature single-celled plants. Being photosynthetic they needed lots of sunlight in order to grow, more than a thick cap of sea ice could have transmitted, so their abundance in sediment core samples from that time period suggests open-water conditions. Furthermore, modern versions of those algae now live in mildly brackish or freshwater habitats, so they also suggest that the polar ocean was diluted by unfrozen, freely running rivers. The additional confining presence of a land bridge across the Bering Strait would have surrounded that low-salinity ocean by land on all but the North Atlantic side of the basin, thereby allowing even more runoff from the encircling continents to dilute it further. Together, these findings evoke an open-water body that resembled the nearly landlocked Baltic Sea of today: saltiest near the Atlantic outlet and least saline at the far inland end.

The South Pole was a very different place back then, as well. There's no firm evidence of large ice sheets anywhere on Earth during the PETM, and no telltale pebbles were sprinkled by drifting, dirt-caked icebergs into the marine muds off the Antarctic coast. Instead, distinctive clay minerals in those muds tell of deep weathering and erosion of soils by warm, wet climates. Kaolinite clay, for example, doesn't often form near the poles today, but it's common in hot, rainy places such as the tropical Niger Delta, and it evidently clouded the southernmost seas 55 million years ago. If ice did exist on Antarctica at all, then it was probably restricted to the highest peaks of the interior.

The deepest layers of the oceans, now among the world's coldest waters, warmed by 7°F or more (4°C) and killed many of the cold-loving creatures that had been living on the bottom. Some investigators attribute this to heating at one or both of the poles because icy temperatures there normally make dense, oxygen-rich seawater sink and spread across the floors of the ocean basins, and they suspect that high PETM temperatures might have prevented that flow. Others believe that heavy rains capped the circumpolar seas with buoyant, low-salinity layers that slowed or stopped the sinking of cool, well-oxygenated surface water.

Whatever its cause, a choking off of life-giving bottom currents apparently smothered deep-sea communities with warm, stagnant, oxygen-poor water. Up to half of all bottom-dwelling foram species vanished from the fossil record, as did other organisms that lived in similarly deep habitats. In contrast, most surface dwellers survived the PETM with no apparent difficulty. But that makes sense, because floaters near the sea surface were probably used to warmer temperatures anyway, and because oxygen could have reached them from the overlying air and from photosynthetic algae floating among them.

As if sweltering and suffocating weren't enough of a burden for deep-ocean life, the water also became corrosive enough to burn that aforementioned red clay band into the sediment record. As we await the life-or-death sentences of many marine species that face carbonic acidification in today's oceans, the PETM example may be painfully instructive. If we go down a similar path during the Anthropocene, will everything from clams to corals vanish with a fizz?

The good news is that many species somehow came through the ordeal intact, especially if they lived in shallow habitats. Some forams died out or developed thinner shells, but others thrived. Oysters and many other mollusks survived, as did quite a few corals. Some chalky planktonic algae even managed to gain shell weight rather than lose it to dissolution. Overall, extinctions in the upper layers of the oceans were largely balanced by the appearance of new species, often with little indication that acidity was the cause of the die-offs. We don't know how so many shell-bearers resisted the acid bath, but it's clear that they did. The bad news is that few, if any, of our modern species existed in their present form 55 million years ago, so we can't say that our marine neighbors have ever survived such severe environmental changes before.

The PETM greenhouse wasn't limited to the oceans, of course, though most of our geological records of it come from marine deposits. To learn what happened on land we turn to terrestrial fossils, especially botanical ones. Many plants back then were remarkably similar to modern forms, more so at least than the animals were, so we can easily identify them from the leaves, wood, and seeds that they left behind, and experts can therefore use them to infer a great deal about past climates.

Fossils exposed along Antarctica's coastline show that it was green with *Nothofagus,* the beech trees that now thrive in temperate rain forests of the Southern Hemisphere, and tall deciduous conifers covered much of Arctic Canada. Such finds speak loud and clear; it was obviously warm and moist enough to turn the polar regions lush and leafy. In North America, many plant species shifted their ranges poleward by more than a thousand miles during the PETM, a mass resettlement that would be like relocating an Alabama flower garden to Hudson Bay in northern Canada.

A research team led by Scott Wing, a paleobotanist at the Smithsonian Museum of Natural History, recently sampled PETM-aged floodplain deposits in Wyoming that were loaded with well-preserved leaf fossils from early relatives of poinsettia, pawpaw, and other southern taxa that had migrated well north of their former ranges. But Wing and his colleagues also sought more subtle details from those ancient remains, as well. What drew their attention most strongly were the shapes of the leaves themselves.

The margins of the leaves fluttering in Wyoming's forests during the PETM were generally smoother and less toothy than they were just before and after it. In most plants, leaf edges are busy centers of the gas-exchange and light-capturing processes that drive photosynthesis. Growing seasons at cool, high latitudes are shorter than those close to the equator, so short-season plants often have to work overtime to get their photosynthetic business finished before winter shuts them down. As a result, the leaves of many high-latitude plants have toothy edges that increase their marginal areas—their "coastlines," if you will—and thereby boost their productivity. Think of serrated beech, maple, and oak leaves in New England versus smooth-margined, tongue-shaped magnolia blades down in Louisiana. Applying a "leaf-margin index" to their collections of fossil foliage, Wing's group estimated that Wyoming's Bighorn Basin warmed by 9°F (5°C), right in line with the marine story: more warming than the tropics felt and less than the poles got.

But how do we know that this seemingly coherent tale wasn't warped by the tectonic wanderings of continents? Perhaps fossils and cores tell of warming because the plates under them shifted closer to the equator back then.

In fact, continental drift is too slow for that. The Atlantic Ocean was a bit narrower back then, but most of its spreading motions run east-west and therefore don't affect latitudinal positions much. The Isthmus of Panama was partially open, which allowed some flowthrough of warm currents that dry land now blocks, and India had yet to plow the Himalayas up into their present skyscraping form. But apart from low areas being inundated by the highest sea levels possible on a defrosted planet, most PETM geography more or less resembled today's layout, at least in a general sense. As a result, we can be sure that the presence of forests in the Arctic means that the far north was indeed remarkably warm, not that it had drifted into the tropics.

It may be difficult to imagine forests growing on what are now stark tundra and polar snowscapes; what were they like? A former student of mine, Chris Williams of Franklin and Marshall College in Pennsylvania, has become a world authority on trees of the early Cenozoic, and his research paints vivid pictures of the past. One of his study sites is the Iceberg Bay Formation on Ellesmere Island in the Canadian Arctic, a storehouse of geologic information whose name and present climatic setting stand in stark contrast to the warm habitats it represents.

"The preservation there is so good that the trees aren't even fully fossilized," Chris explained recently. "It's more accurate to say that they're mummified instead." The original cellulose is gone after 55 million years, but enough cell structure still remains that many of the leaves, branches, and trunks look almost fresh. "The Arctic forests were full of deciduous dawn redwoods, some of them a hundred feet tall, and they covered northern Canada and Alaska for tens of millions of years during the early Cenozoic. You could also find lots of ginkgos and cypress there, which makes sense considering that these species like it wet, and ice-free river corridors and swamps were common in the Arctic back then."

Most of the trees dropped their needles or leaves every year, not necessarily in response to seasonal cold but more likely to the long darkness of high-latitude winter. However, their absence up there today does reflect long-term Cenozoic cooling that drove the trees south into regions where other species eventually outcompeted them; only a few, if any, undomesticated dawn redwoods and ginkgos survive now in the wild, apparently only in China. And during the last 2 to 3 million years, ice ages finished off whatever remained of those high-Arctic forests.

With woodlands thriving in the polar regions, animals spread widely and diversified as they do in today's tropics, but some of the richest sources of information about those creatures are located in the dry hill country of northern Wyoming and central Utah, parts of France and Belgium, and China's Hunan Province, where fossil-laden sedimentary rocks are well exposed for sampling. PETM-aged strata paint distinctive red and purple stripes along the exposed flanks of many such formations, luring paleontologists to generous troves of primitive teeth and bones.

Mammal remains in such deposits tend to draw more attention than those of other animals do, partly because there are lots of them to unearth, but also because the PETM marks the noteworthy first appearances of entire groups—"orders," in the taxonomic sense. This gives it a special significance for us, being mammals and all. A new Eocene order of cloven-hoofed mammals, for example, gave rise to deer and cattle. Another order with solid hooves later spawned modern horses. And a new lineage of big-eyed, large-brained, lemurlike creatures eventually branched out into those of monkeys, apes, and humans.

But although they help to draw new boughs on the mammalian family tree, those long-extinct animals don't tell us as much about PETM climates as the plants do because we know little about them other than what their hard parts reveal. In fact, most of us wouldn't know what to make of them if they showed up in our backyards. Back in the early Cenozoic, the ancestors of horses were poodle-sized, hunch-backed omnivorous things, and protowhales walked on four legs. Snaggle-toothed, long-jawed condylarth predators galloped after their prey on toes tipped with tiny hooves; some scientists half-jokingly refer to them as "wolf-sheep." And then there were the hyaenodontids, *Labidolemurs,* and *Macrocraniums,* whose very names evoke the exotic fringes of biological fantasy.

This unfamiliarity factor hinders our ability to use such creatures as temperature indicators. Fossils of tapirlike animals that were recently found on Ellesmere Island appear to make sense in the context of climatic warming because a modern tapir is rather like a floppy-snouted pig that grunts and dashes about in neotropical rain forests. However, tens of millions of years separate the PETM species from those of today, so these animals weren't necessarily tapirs as we know them now. Putting such a familiar name on them might therefore mislead us in our climatic sleuthing.

A species like woolly mammoths, which appeared much later in the Cenozoic, can better represent past climates because we've found their wool as well as their bones. Unfortunately, most PETM fossils are of the strictly bony type. Imagine that we had nothing but bones left to tell us about mammoths, with no clue about the Big Hair that survives in permafrosted specimens and that so clearly links them to cold weather. An unfleshed mammoth skeleton looks much like that of a modern elephant, and a scientist excavating ice age mammoth bones in Europe might falsely conclude that it was warm enough back then for "elephants" to migrate up from Africa, when in fact the great beasts endured frigid conditions every winter. That's the situation we face with the Ellesmere "tapirs." With only skeletons to go on, we couldn't be sure that the PETM tapirs liked it hot or cold if we didn't also know that Ellesmere Island and other polar regions were cloaked in forest rather than the ice and tundra that we find there today.

Another possible climatic clue is that most of the PETM mammals were oddly small, about half the size of their counterparts in older and younger deposits. Some paleontologists have attributed this to warmth because smaller bodies resist overheating better than bigger ones do; the inner furnaces of very large animals can sometimes produce more body heat than their skin easily releases into an already-warm atmosphere. But others counter that elephants, rhinos and giraffes do just fine in tropical Africa today, and that huge dinosaurs thrived during the warm Mesozoic era.

A more recent hypothesis suggests that greenhouse gases caused the stunting by altering the nutritional contents of the plants upon which the animals grazed. The sharp dip in body sizes during the PETM roughly parallels the dip in ^{13}C concentrations caused by the greenhouse gas surge, and high concentrations of CO_2 are to many plants as candy is to kids—tasty but not very nourishing. They can make vegetation grow rapidly, but the bodies of plants grown under such conditions tend to run low on other nutritional substances such as nitrogen, rather like a building made of cheap, easily assembled cardboard rather than concrete and steel. In the case of animals, one possible response to such an unwholesome menu is to grow less, and that may be what the PETM mammals did. Some experts worry that future plants and animals might also do the same in a super-greenhouse future.

Evidence for just such a reduction in the food value of plants may be revealed in a collection of leaf fossils excavated from the Bighorn Basin of Wyoming. A study led by geologist Ellen Currano, another former student of mine who now holds a faculty post at Miami University, Ohio, found more signs of piercing, tunneling, and chewing in leaves that were deposited during the PETM than in slightly younger or older layers. "We think that insect populations were more diverse at the study site back then because many species migrated in from lower latitudes with the warmer weather," she explained. "And another factor could be that the leaves became less nutritious because of the high carbon dioxide concentrations, so the insects had to eat more in order to get the same amount of nutrition from their food." When Currano's team described their findings in a recent issue of the *Proceedings of the National Academy of Sciences*, they ended their article with a warning that future greenhouse gas buildups might have similar consequences.

Whatever the reason for their small body sizes, those early Eocene mammals didn't seem to mind the weather. Though some species went extinct, including the long-tailed, "almost primate" *Plesiadapis* and the long-jawed, "sort of crocodile" *Champsosaurus*, many others appeared or persisted all the way through the long thermal excursion. PETM species seem to have lived pretty much everywhere because the global greenhouse more or less leveled the pole-to-equator temperature differences that now isolate penguins from pandas. Animals wandered more widely than usual in that warmer world, and many that invaded the Arctic also crossed the Asia-America land bridge into vast new territories. When an early primate appeared in China, for example, it dispersed into North

America so quickly that it almost seemed to have appeared in both places at once.

We aren't sure exactly how rapidly the PETM's dramatic environmental and evolutionary changes happened because our methods of dating very old sediments don't allow for perfect precision. If they were annually layered, all nicely stacked like sequential pages in some ancient ledger, then we could simply count the laminations across the transition zone to tell exactly how long it took to reach full heat or how long the cooling tail of the PETM's CO_2 curve was. But most deposits aren't like that, so we have to use relatively blunt methods to get at their ages. Carbon-14 doesn't work on such ancient records because its atomic clock runs down fairly rapidly, and longer-lasting tools, like radioactive potassium or uranium, don't clearly register the short time increments that characterize truly rapid or brief events.

However, such concerns are minor. There are plenty of lessons in the PETM example that can be applied to our modern world. For starters, it demonstrates conclusively that a super-greenhouse is not just some doomsayer's morbid dream; it really can happen.

We don't know exactly how it began, but we do know that its most notable effects lingered for roughly 170,000 years, which is within the ballpark of what most farsighted computer models are predicting for a relatively extreme-emissions scenario.

We don't know exactly how fast it began, but we do know that it happened abruptly in geologic terms, reaching peak warmth over the course of centuries. In that sense, our own climb toward thermal maximum in modern times is somewhat similar to the carbon-rich path that once led the world into PETM hothouse conditions, but the slope of our climb is apparently even steeper than that of the earlier one.

We don't know exactly how much the world warmed back then, but we do know that global average heating on the order of 10°F or more (5 to 6°C) shrank the temperature differences between high and low latitudes, turned the Arctic Ocean into a brackish lake, erased the last large frozen habitats from the continents, and lifted sea levels to their highest possible positions. Similar environmental disruptions are therefore also within the range of possibility for a comparably extreme carbon emissions scenario in Anthropocene times.

We don't know exactly how much carbon-bearing gas was required to trigger the PETM, but we do know that it didn't necessarily appear all at once. Positive feedback loops almost certainly played a major role, and they could just as well amplify a modern fossil fuel emission into a supergreenhouse like that of 55 million years ago. Unfortunately, we don't know exactly what critical threshold of temperature or CO_2 concentrations would trigger those feedbacks, and we can only hope that a relatively moderate emissions scenario doesn't exceed it as well.

We don't know exactly how high greenhouse gas concentrations rose, but we do know that they warmed and acidified the deep sea enough to devastate bottom-dwelling communities and to burn a red layer into the ocean floor. Sediment cores suggest that it took thousands of years for the worst of the acidification to subside. If we take a 5,000-Gton emissions path into the future, then we're likely to leave a similar scarlet mark of shame on future sediment records.

We're not sure exactly how the high CO_2 concentrations themselves affected biota, but they might have reduced the nutritional value of plants, stunted the growth of mammals, and encouraged herbivorous insects to attack vegetation more vigorously. Any such effects on future crops, herds, or wild communities would be most unwelcome.

The animals that lived through the PETM weren't exactly the same as those of today, but we can still learn much from the lessons that they offer. Many species thrived in those warm climates, showing that the higher temperatures weren't necessarily a hindrance to life on land. But we also see that timely migration was an important key to biological success under those conditions.

We can only guess what it might mean to force a PETM-style climate structure onto today's complex and unpredictable network of nations. Surely there would be winners as well as losers in any such shift. But the international squabbles that have recently erupted over the Arctic's newly opening sea lanes would pale in comparison to the territorial conflicts that might follow the de-icing of a mineral-rich continent on the South Pole, not to mention the displacements driven by an eventual 230-foot (70-m) sea-level rise or extreme derangements of today's climatic zones.

On a relatively bright note, we also know that many plants and animals, including our own early primate ancestors, made it through the

PETM just fine. If we could time-travel back to the early Eocene, many of us might even find its climates quite pleasant to live in as long as we could also get used to experiencing them above the Arctic Circle in the company of faux tapirs and snarling wolf-sheep. After all, paleontologists label the more prolonged hot phase that followed the PETM with the positive-sounding "early Eocene climatic *optimum*," rather than something more akin to "climatic disaster," because life in general seems to have thrived in it.

But we also know that the survivors of the PETM passed through those earlier warm times without human activity working against them. If we launch a new superhothouse of our own, the descendants of those Eocene ancestors will find it difficult to follow us into the future. No longer free to colonize new land- and seascapes as their preferred climatic zones shift poleward, the cumulative wealth of modern biodiversity could melt away along with the great ice sheets in this, the carbon-enriched Anthropocene.

5

Future Fossils

I, unfortunately, was born at the wrong end of time,
and I have to live backwards from in front.
—*Merlyn,* The Once and Future King

One thing about the greenhouse effect that makes it so difficult for some of us to take seriously is that it's invisible. Carbon dioxide leaves no stain on the air, no odor betrays its presence, and most of its climatic effects are subtle enough to blend into a camouflaging background of natural weather fluctuations. But for many scientists who make their living by studying invisible things, fossil fuel carbon has a decidedly tangible presence. To them, it's not just an abstract concept that vaguely influences future generations; it's a demonstrable fact of modern life, something to be remembered constantly and added to the list of routine adjustments made in the course of a day's work. This little-known aspect of carbon pollution is even more difficult for most of us to detect than a slow rise in global average temperature, but it's actually even more pervasive because it permeates our bodies as well as the air, water, and sediments around us.

For professionals who deal directly with the geochemical effects of worldwide fossil carbon contamination on a regular basis, any doubt about its existence could only be found among the uninformed or the most hardheaded contrarians. Although its best-known attribute is its ability to change climate, the pollution itself isn't necessarily all bad in this context; some aspects of it, as we'll see in this chapter, are arguably somewhat positive. But its undeniable presence makes a powerful statement: the massive human influence on global carbon chemistry is real, measurable, and scientifically significant.

Chief among those for whom fossil fuel pollution is a decidedly here-and-now issue are three kinds of specialists: ecologists who monitor the levels of biological activity in ecosystems, forensic scientists who seek clues about when and where things have happened in the relatively recent past, and geoscientists who read deeper histories in ancient deposits of mud, ice, wood, and stone. These investigators use carbon atoms as tools in support of esoteric techniques that have revolutionized scientific understanding of the natural world, but their work is now strongly influenced by the same gases that are also changing climates. Fossil fuel carbon pollutes far more than air and water alone; it also contaminates the fundamental atomic structures of organisms and ecosystems, changing them in ways that most of us are unaware of.

Some brief introductions are in order here. In our daily lives we deal with chemical elements only as homogeneous clumps of unimaginably numerous particles. But not all carbon is alike, as the diagnostic ^{13}C signal in PETM-age deposits makes clear. Scientists distinguish among three forms of it, separating them on the basis of weight, and each plays a unique role in the Anthropocene world.

These isotopes—or varieties—of carbon are like fraternal triplets, and apart from their weight they are similar enough to one another to appear in the same kinds of molecular compounds, from CO_2 and methane to proteins and genes, and they participate in many of the same chemical processes. They differ mainly in the number of tightly clustered particles that make up their central nuclei.

Carbon-12 (or ^{12}C) is what you might call the "normal" sibling, and 99 percent of all carbon atoms are of this sort. However, one in every hundred or so carries an extra neutron in its nucleus, which raises its atomic weight and changes its name to carbon-13 (^{13}C). Beyond that the difference is minimal, rather like a slight paunch on a formerly slim, middle-aged person.

And then there's carbon-14 (^{14}C). One might say that it's the bad apple in the bunch, a troublemaker with a reputation for making sudden outbursts. It is much rarer than ^{13}C is, and one neutron heavier. Furthermore, it's much younger than its siblings are. The two lighter isotopes were born long ago in the internal fusion reactors of distant stars, and they are essentially immortal, but ^{14}C forms continuously in Earth's upper atmosphere and it is doomed to die an early death. Its explosive personality stems from the physical stresses induced by that extra neutron, which doesn't fit very well in the already-crowded nucleus; it's almost as if its neighbor particles want to flick the superfluous neutron away. Because of this instability ^{14}C is "radioactive," meaning that sooner or later it spits out a nuclear chunk and undergoes a sudden conversion to a lighter, more stable state of existence (as a born-again nitrogen atom, to be precise).

A gram of carbon from coal, oil, or natural gas contains less ^{13}C and ^{14}C than a gram of carbon in atmospheric CO_2 does. In part, this is because the long-ago plants and algae whose bodies formed those fossil fuels selected normal ^{12}C over the heavier isotopes when building the molecular frameworks of their cells, just as their descendants do today. When a plant or alga inhales a CO_2 molecule, it treats it like a seafood chef treats a hard shell clam. Out comes the shucking knife and off come one or both of the shells to expose the nutritious gob at the center; in the case of plants, the clamshells are twinned oxygen atoms and the gob is a carbon.

But like a finicky connoisseur, a leaf or algal cell is choosy about the molecules it consumes. Carbon dioxide molecules that contain heavy ^{13}C and ^{14}C atoms tend to be tossed aside in favor of the lighter, normal versions. Nevertheless, a very few oddball carbons do manage to slip into the mix and take up residence in the living tissues of plants.

Having already passed through the filters of once-living bodies, the carbon stored in fossil fuels has been partially scrubbed of those heavy isotopes. Furthermore, any unstable radioactive ^{14}C that once existed in those ancient deposits has long since broken down. As a result of these processes, fossil fuels contain only a tiny remnant of the isotopic carbon diversity that one might encounter in a random puff of wind. And so do the carbon-rich fumes that emerge when fossil fuels are burned. In an odd twist of chemical ecology, fossil fuel exhaust actually helps to "purify" the air of heavy isotopes by diluting it with lightweight ^{12}C.

The human-driven deficit of ^{13}C and ^{14}C in the atmosphere is called the Suess effect, after Austrian American scientist Hans Suess who originally measured it (and who, to his irritation, was often the recipient of fan mail to his contemporary, "Dr. Seuss"). Because atmospheric carbon flows abundantly through the world's food chains, moving from plant sap to rabbit muscle to fox DNA, the telltale signs of the Suess effect are intimately woven into the molecular tapestry of life on Earth. Normally, they go unnoticed, though we carry them in every lump and fiber of our bodies. But for many ecologists, the Suess effect is a threat to the accuracy of one's data.

Imagine investigating the ecological history of a lake, as was done recently for Lake Erie. Phosphorus pollution from cities, lawns, sewer pipes, and farms fed nuisance algae living in its surface waters for many years, boosting nasty aquatic scum growth just as fertilizers boost crop yields on land. Each year, new layers of dead algae piled up on the lake floor, enriching the mud with their carbon-laden remains and preserving a nicely layered, if rather fetid sedimentary record of the lake's pollution problems.

Like plants, algae preferentially remove lightweight ^{12}C from solution as they absorb waterborne CO_2 for photosynthesis, and the green, plankton-choked surface of Lake Erie began to show signs of that selectivity. Like a desktop candy dish full of tasty jelly beans plus a few pebbles, over time the water became more and more abundantly stocked with unwanted heavy CO_2 molecules that the algae left behind. As the chance of accidentally plucking a pebble of heavy ^{13}C rather than a desirable ^{12}C increased, generations of green slime became more and more enriched with ^{13}C as they lived, died, and sank to the mucky bottom of the lake.

In the mid-1980s, after strict water quality regulations were put into place, phosphorus inputs to Lake Erie decreased to a quarter of their earlier volume; the city of Detroit alone reduced its annual phosphorus outflow by nearly two-thirds. But was the strategy working?

University of Florida ecologists Claire Schelske and David Hodell decided to find out by probing the sediment archives under the lake. They knew that the ^{13}C contents of bottom muds should have increased while the algae were abundant. They also knew that cleanup efforts should have shrunk the algal crops and therefore slowed the delivery of ^{13}C to the bottom. By lowering weighted core pipes to the lake floor, they collected sedimentary records of the last century and measured the ^{13}C contents of sequentially stacked layers of mud. Sure enough, the

^{13}C values rose to a peak in strata that were deposited during the late 1960s, when the ^{13}C-enriched algae were at their thickest. Then the trend reversed direction, dropping all the way back to pre-pollution conditions by the 1980s.

That seemed like good news at first, but something about it didn't ring true. The landscapes around Lake Erie are now more heavily populated and developed than they were at the turn of the twentieth century, and yet the ^{13}C records seemed to show a complete recovery. Were the pollution controls really so effective? Schelske and Hodell doubted it, and they had a good idea of what might cause such an illusion of exaggerated success.

Sure enough, it was the Suess effect at work. Carbon-13 concentrations in algae worldwide have been declining throughout the twentieth century, thanks to the burning of fossil fuels. When that factor was included in the sediment core analyses, the recovery of Lake Erie didn't look nearly as complete as it had before. The hoped-for turnaround to cleaner waters did occur, but the Suess effect had exaggerated the subsequent decline of ^{13}C concentrations in the algae-laden muds. In fact, the lake was still a fair bit slimier than it used to be in the early 1900s.

The newly corrected take-home message was still supportive of regulations; they had certainly saved Lake Erie from becoming a vat of green pea soup. But by recognizing the distorting effects of fossil carbon in the environmental record, Schelske and Hodell also showed that more cleanup work still remains to be done.

In like manner, fossil carbon pollution is also complicating studies of climatic change and aquatic ecology elsewhere around the world. In 2003, I worked in the lakeside town of Kigoma, Tanzania, as an instructor for the Nyanza Project, an undergraduate research training program supported by the National Science Foundation and the University of Arizona. Shortly after my arrival, several of my coinstructors and colleagues published papers showing that Tanganyika, the lake upon which our field studies were based, is warming.

Lake Tanganyika is huge, roughly 400 miles long (670 km) and almost a mile deep (1,470 m), making it the world's second deepest lake, after Siberia's Baikal. Most of it is devoid of animal life because the deeper portions contain no oxygen; only the upper 300 feet (100 m) of surface water has enough photosynthetic algae and churning wave action to oxygenate it. Nonetheless, hundreds of species of colorful cichlid fish live in that relatively thin upper zone and nowhere else on Earth. The lake's beautifully clear waters also support delicately sculpted snails, freshwater crabs and jellyfish, and even aquatic cobras, all unique to Lake Tanganyika.

The two sets of authors, headed by Vassar College's Catherine O'Reilly and Piet Verburg from the University of Waterloo, Ontario, compiled different assortments of data and observations, but they independently reached the same conclusion. The waters of Lake Tanganyika have warmed by about 2°F (1°C) during the last century, a pattern that resembles similar trends in neighboring Lakes Victoria and Malawi, as well.

The O'Reilly team went another step further, though, by linking that warming to the main protein source of many local Tanzanians: fish. By studying Tanganyika sediment cores, they found that the younger layers contained less ^{13}C than the older layers did, which could mean that algae were declining as they did during the cleanup process in Lake Erie. But

Students and staff of the Nyanza Project preparing to collect a sediment core from Lake Tanganyika in 2003. *Curt Stager*

nobody has been controlling phosphorus pollution in this lake, so something else must be at work.

O'Reilly's group proposed that warming had stabilized the lake's surface by making it less dense and therefore more buoyant, rather like capping it with a thick layer of oil. Planktonic algae can find it difficult to stay afloat in the increasingly isolated sunlit zone, and they begin to sink into deeper, darker waters where they die of light deprivation, as plants will blanch when locked away in a darkened room. Such a planktonic population decline, in turn, could weaken the entire food web and leave the sardinelike fishes of the open lake with less to eat. According to O'Reilly's calculations, this might cause large declines in annual catches, quite sobering news in an economically depressed region where about a third of all dietary protein comes from Tanganyika fish.

However, the case may not be as neatly closed as it first appeared to be. In a later reevaluation of the evidence for algal declines, Verburg applied a standard Suess effect correction to the sedimentary records of ^{13}C, the O'Reilly team's measure of plankton productivity. When he included the skewing effects of fossil carbon in the analysis, he found that ^{13}C concentrations (that is, algal growth) may actually have increased where agriculture and other human activities pollute the near-shore waters with nutrient-laced runoff. Although the lake is indeed warming, it is now unclear what effect, if any, that trend is having on the resident plankton or fish.

These and a growing number of similar stories show that fossil carbon pollution contaminates ecosystems worldwide. Scientists usually measure the Suess effect in units that reflect the abundance of heavy carbon in relation to regular carbon-12, and the values of "delta-^{13}C" (or ^{13}C divided by ^{12}C) have been falling faster and faster as we pump more and more lightweight fossil carbon into the atmosphere. During the eighteenth century, a typical sample of CO_2 from the air would yield a delta-^{13}C value of -6.3 parts per thousand (ppt); today, after two centuries of dilution with ancient carbon fumes, it is down closer to -8 ppt.

Though few of us have noticed it, we're already living through a global ^{13}C decline comparable to the one that the PETM super-greenhouse caused 55 million years ago, and most of it has happened during the last century. The Suess effect now lowers global delta-^{13}C values by nearly a fifth of a unit per decade. At that rate, our Anthropocene isotope excursion could reach that of the PETM by the time airborne CO_2 concentrations peak within the next few centuries. If a team of oceanographers drives a core pipe into sediments of the deep sea at some point in the distant future, they'll find signals from our carbon-enriched times that resemble those left by the PETM in the reddened layers of Lowell Stott's marine cores. The physical effects of these isotopic changes on our daily lives are easily missed, but to scientists who work closely with them their historical significance is breathtaking. Without even knowing it, we're writing our own environmental epitaphs in indelible carbon isotope ink.

Dating the records of our times won't be easy, though. The same Suess effect that dilutes ^{13}C in the environment today also dilutes ^{14}C, the radioactive tool of choice for dating old historical objects. Carbon-14 is not a perfect guide to the past; its shelf life (more properly termed "half-life") is short in comparison with those of radioactive marathon runners such as uranium-238, which is used for dating rocks that are billions of years old. It only works on things that contain carbon, such as wood, bone, shell, peat, or aquatic muds, and only on those that are also less than about 50,000 years old. Even so, the time-tracking ability of radiocarbon dating has revolutionized our understanding of human and environmental history. Unfortunately, it won't be quite as useful for future scientists, thanks to the Suess effect.

It's easy to imagine some of the questions that scholars might want to ask when they look back on the story of our times. When did the last ice sheets melt away? At what point did the oceans finish acidifying? How rapidly were national boundaries reshuffled in response to sea-level rise? Some of those questions will be answered by written documents, but not necessarily as many as we might think.

Much of today's recorded history will eventually be lost simply because so many of its documents are electronic. The devices and codes that create and decipher them change so often in the interest of big business that they become useless within decades or less, not to mention centuries. The effects of this are already apparent in my own home. I'm still saving some old-fashioned floppy discs that used to feed data into my

TRS-80 computer back in the 1980s, even though I'll never be able to read them again. They're so full of hard-won information that I just can't bear to throw them away. And don't even mention my old eight-track tapes, or the memory chips from my first digital camera that the latest card readers won't read.

Fortunately, the geological archives of ice, coral, tree ring, stalagmite, and sediment layers around the world are still depositing their annual environmental updates as they have for millions of years. Even deposits in what are now thinly inhabited regions contain indelible signs of our presence on the planet, and carbon isotope glitches or buildups of airborne lead and other pollutants will identify them as such. Those signals will speak as clearly as written text to knowledgeable people in later phases of the Anthropocene, long after today's information technology moguls and their ephemeral media are forgotten.

Or rather, all but one of those traditional geochemical signals will continue to come through clearly. As a direct result of our fossil fuel emissions, the code with which we read ^{14}C records has now been scrambled. Before I can explain exactly why this is so, you'll need to understand how ^{14}C dating works. Let me begin by informing you, dear reader, that you are radioactive.

Because the carbon-based food molecules that we build our skin, flesh, and bones from contain small amounts of ^{14}C, we are all slightly radioactive. And that's because the photosynthetic plants and microbes that support Earth's ecosystems absorb ^{14}C atoms with the CO_2 that they inhale day after day. Radioactive carbons in your average lungful of air are quite rare, forming less than one in a trillion CO_2 molecules, but that slight dose of natural radioactivity infects all of us through the global network of food webs. As much as one in every four cells in your body contains a ^{14}C atom in its DNA or in the histone proteins that surround it, and a gram of carbon purified from any given body part contains enough ^{14}C to trigger a little more than a dozen telltale clicks per minute on a Geiger counter, a frequency roughly equivalent to that at which an average person draws breath. According to one estimate, a typical adult human experiences about 300 internal ^{14}C explosions every second.

How did the air become radioactive in the first place? Cosmic rays. They come at us from all directions, in waves of subatomic debris that blast through space from distant stars and galaxies. After traveling many billions of miles on any one of countless possible paths, some strike the upper reaches of our atmosphere and collide with air molecules there. The impact of a fast-moving cosmic neutron striking a nitrogen atom, the most common component of air, can kick a tiny proton pellet out of the atom's inner kernel, the nucleus. By the rules of atomic nomenclature, it is now no longer a nitrogen atom; it has become a radioactive carbon-14 atom.

Sooner or later, the new ^{14}C atom loses control of its unstable core. The atom flicks out a tiny nuclear beta particle and reverts back to plain old nitrogen again. That's what radioactivity is all about: the forceful ejection of debris from overweight atomic nuclei.

In some cases, the flying chunks that radioactivity sprays hither and yon are energetic enough to injure living things. Radium, for instance, emits a continuous volley of potentially dangerous subatomic buckshot. We are well advised to stand clear of such powerful nuclear radiation sources because their invisible ejecta can plow through our cells and cause burns, radiation sickness, or cancer. But carbon 14 is merely a pop-gun by comparison, and under most circumstances its shrapnel is unlikely to damage your genes or tissues. The only way it can hurt you is if you inhale or swallow it so it can move in close to your cells by way of your bloodstream. Of course, we do exactly that every time we breathe, eat, or drink. In other words, it's not the ^{14}C around us that poses a threat; it's the ^{14}C within us.

In the case of ^{14}C, we face several possible health hazards. It lies deeply embedded in our carbon-rich body tissues, so when any given bit of it explodes the recoil or the flying pellet may damage an adjacent molecule or cellular feature of importance. But there's yet another threat as well. Carbon is a key component of our genes, so some of the ladderlike chains of DNA that they consist of contain radioactive time bombs. When one of them goes off, the precisely organized structure of the gene changes in ways that may alter the health or behavior of its host cell. And if one or more of the supporting structural proteins that bind to it fly apart, it could change how a gene folds or unfolds as it is being stored or opened for use. On a really bad-luck day, such mutations could cause a birth defect or a deadly tumor.

The risk is real, though blessedly small because of our built-in cellular repair mechanisms, and until recently it was thought to be an unavoidable fact of life. But no longer. The boundlessly innovative energy of the marketplace has spawned a way to make money from this ancient bane of human health. Why settle for organic, pesticide-free produce when you can buy it radiocarbon-free as well? All that is needed is to raise vegetables in a tightly sealed chamber that allows the composition of the air inside to be controlled. The exact details of the method are a trade secret because, according to one patent application that I found online, "they're so simple and obvious," but they seem to have something to do with burning coal, oil, or gas and running the CO_2-charged exhaust into the greenhouse chambers. Fossil fuels are devoid of ^{14}C, so when crop plants inhale those vapors they grow up free of radioactivity, probably the first organisms ever to do so on this planet.

Most ^{14}C atoms form rather quietly in the upper atmosphere, but they don't go unnoticed there for long. Oxygen atoms, the second most common component of air, quickly locate solo carbon atoms and stick to them. Within hours of their birth, freshly made ^{14}C atoms and their oxygen hitchhikers drift away on the wind as carbon dioxide. Eventually, those radioactive CO_2 molecules mix downward into the lower reaches of the atmosphere, where they can be sucked into the cells of photosynthetic bacteria, algae, and plants. There they blend in with the normal carbons until, sooner or later, they revert back to their original nitrogenous selves in a burst of energy and flying particles.

We, in turn, eat radioactive animals and plants and use the nutritious components of their tissues to build our own bodies. We thereby divert some of the world's biological carbon flow into our bodies with each meal and release some with each exhalation, excretion, secretion, exfoliation, and birth. But when we die our bodies are cut off from the stream and, from that moment on, we stop replacing the ^{14}C atoms that break down inside us. After 5,740 years, about half of our ^{14}C load will have vanished. As another 5,740 years pass, half of those remaining atoms break down, and so on until all of them are gone or they're too rare to measure accurately. Normally, that near-total decline takes about 50,000 years.

Much to the delight of scientists, the average rate of ^{14}C decay is so reliably stable that we can use it as a molecular clock to date once-living things. To do that, we need only measure how much of it remains intact within an object. If about half of the expected amount of ^{14}C is present, then the substance is probably close to 5,740 years old. If only a quarter remains, then it's twice as old, and so on. The less ^{14}C you find in something, the older it's supposed to be.

Unfortunately, the opening centuries of the Anthropocene have now thrown that system into disarray. The fossil fuels that we burn are millions of years old, so the unstable ^{14}C atoms that were originally present in them broke down long ago. The CO_2 that forms when coal, oil, and natural gas are burned is a "dead" gas, unlike the stuff that goes up our chimneys when we burn recently cut ^{14}C-laden wood, so by recycling tons of stable fossil carbon back into circulation, we've reduced the natural radioactivity of the atmosphere. At last, a bright side to air pollution; this kind slightly dilutes the radiocarbon load in the air and in our bodies.

If future historians use radiocarbon dating on objects from these early stages of the Anthropocene, they'll face a serious problem. All of the bone, hair, wood, aquatic mud, and other such materials that formed between the late 1800s and mid-1900s were less radioactive than usual because they were diluted by stable fossil ^{12}C in the air. When the maxim of "less ^{14}C means older" is applied to such samples, they yield ages that appear to be artificially old.

According to Darden Hood, of the Beta Analytic dating lab in Miami, the Suess effect on radioactive carbon first became clearly noticeable in the 1890s. "If you analyze tree rings that were laid down between the late 1800s and the 1940s," he explained to me recently, "you find about three percent less carbon-14 in them than you'd otherwise expect."

To put it another way, imagine radiocarbon dating the remains of a U.S. infantryman who died in World War I. Fossil fuel carbon was already contaminating the air, oceans, and human bodies of his day enough to change their apparent ages. The time offset in the soldier's bones would therefore suggest that he died 200 to 300 years earlier, well before the nation that he fought for was born. Although nobody could have known it then (^{14}C wasn't discovered until 1940), everyone who lived during the first few decades of the twentieth century was a living fossil, thanks to global carbon pollution.

But things became even more complicated after we invented yet another kind of carbon pollution during the closing stages of World War II. The United States, the former Soviet Union, and other nations began to test powerful thermonuclear devices high up in the atmosphere. Those detonations were, in a sense, like artificial cosmic ray bursts that created radioactive ^{14}C by smashing nitrogen atoms in the air. Hundreds of nukes exploded in aboveground tests during the 1950s and 1960s before a partial ban outlawed them, collectively producing so much ^{14}C that it temporarily overwhelmed the Suess effect from our fossil fuel emissions. By 1963, at the height of that grim fireworks display, atmospheric ^{14}C concentrations worldwide had nearly doubled.

As a result of that global nuclear contamination, all organisms that have sprouted, crept, flown, or swum since the 1950s have been artificially enriched in radioactive bomb carbon. It travels with you wherever you go, in the skin of your hands, in the pages of this book, in the snuffly wet nose that your dog is pressing against you in hopes of winning a radioactive snack treat. Published estimates of the resultant damage to human health over the subsequent decades are difficult to confirm, but they range from hundreds of thousands to millions of cancers and birth defects.

Not all of the effects of bomb carbon pollution have been bad, though. Forensic investigators have learned to use it in amazingly creative ways.

Imagine that you have just invested a small fortune in a case of fine wine that is supposed to have been vinted in 1910. Why not check the radiocarbon content of the alcohol, just to be sure? If it contains excessive amounts of ^{14}C, then the wine came from grapes that ripened during or after the bomb-happy 1950s. Too bad for you.

Or perhaps you have bought an ivory sculpture that is supposed to have been made from legal tusks that were harvested before an international ivory ban was enacted. If you want to be sure that you're not supporting the elephant poaching trade, you could check to see if the amount of ^{14}C in that sculpture matches the atmospheric concentration of bomb carbon from the supposed harvest year.

Even wildlife biologists have been turning the radiocarbon lemon into lemonade. At Isle Royale National Park, a wooded island in Lake Superior, the concentrations of bomb carbon in the teeth of moose have been used to determine when the animals were born.

And what has all that bomb carbon done to radiocarbon dating? By boosting radioactivity levels everywhere, it has completely reset the planet's radiocarbon clock. Early in the twentieth century, the diluting Suess effect made everything seem to be older than it really was, but nowadays every living thing on Earth has more ^{14}C in it than it otherwise would. When you now apply the traditional formula to modern objects, you don't get artificially ancient radiocarbon ages any more, but you don't get accurate "zero" ages, either. Bomb carbon has pushed the dial on the theoretical isotopic clock so far forward that it runs us right through the present moment and into the future. You may be living in 2011 but most of your body—or, at least, its apparent age—lies many centuries ahead of you on the radiocarbon time line.

According to Darden Hood's calculations, I was seemingly born in 5300 AD. In reality, I entered this world in 1956, but the food molecules that Mom inadvertently passed along to me through my fetal umbilical cord were infested with ^{14}C atoms that formed in mushroom clouds, probably somewhere over the Pacific. The bomb carbon that flooded the food webs of 1956 made me so radioactive that my newborn body lay three virtual millennia ahead of me in the future.

But bomb pollution is fading fast now. It's not that it breaks down very quickly; that process will take thousands of years to put a noticeable dent in the reservoir. It's because carbon-bearing minerals and the bodies of aquatic organisms are falling into layered tombs on the sea floor year after year, a process that locks their radioactive carbons away with them along with a large fraction of our fossil carbon emissions. The decrease is exponential, and most scientists expect the Suess effect to regain dominance over bomb carbon within another decade or two.

Because the bomb carbon effect is growing weaker and weaker, it is also driving our apparent radiocarbon ages less and less deeply into the future. Not only do today's newborns contain far less ^{14}C than I did in 1956; the food we eat brings us ever lower doses of the stuff, too. This should also make our present isotopic age offsets less extreme than the ones we were born with.

I asked Hood to crunch the numbers for me. After a brief pause punctuated by the clicks of a calculator, he answered. "Most people living today still have a fair bit of bomb carbon in their bodies. If you eat the same kind of breakfast cereal that I do each morning, then I would expect your radiocarbon age now to be something like 580 years. In the future."

Somewhat confusing, perhaps, to be living backward through time like Merlin the Arthurian wizard. It's as though I started my life in 5300 AD and ended up in 2589 AD by late middle age. But even this bizarre aspect of the bomb carbon saga has a decidedly uplifting side to it. Biomedical researchers are using the known annual decline of bomb carbon concentrations as a tool to help them answer important and long-standing questions about human bodies and health.

Do our brain cells form only when we're young? If so, then the effects of "sex, drugs, and rock and roll" on our neuron supplies might be more persistent than we would like. Are fat cells permanent fixtures in our hips and bellies? If so, then diets are only temporary holding actions against obesity. And what about that tumor we've just spotted; did it balloon up recently or is it simply a slow-growing lump?

Bomb carbon analysis can help to answer all of these questions. Different parts of our bodies contain different amounts of ^{14}C, not because they reject or absorb it differently but because they formed at different times in our lives. Matching ^{14}C contents to specific years has been used to bring us both good news and bad; the visual and memory neurons that we lose to fun and games are not renewed, sad to say, but our body-fat cells are replaced every eight years or so, and the central cores of many tumors are potentially datable enough to help with the design of cancer treatments.

The time-warping effects of bomb carbon will only influence objects that formed between the 1950s and, say, 2020 AD. But those effects are strange indeed, and future scientists will have to deal with them if they try to probe the artifacts of our times with radiocarbon dating. Of course, this is among the least of our worries as we consider the consequences of Anthropocene carbon pollution; the disruption of radiocarbon ages, per se, isn't going to melt any ice or drive any species to extinction, and the dilution of ^{14}C with inert fossil carbon is, if anything, a health benefit.

But what if we look at it another way? Most of us hope to leave some sort of positive mark on the world after we're gone, some sign that we've been here. It's not much to ask for, even if the urge may be rooted in ego gratification or a deep-seated fear of mortality. However, by ruining the radiocarbon age labels that will accompany the remains of our bodies and our civilizations down through the ages, we have utterly scrambled the chronological filing system of history.

Future scientists who hunt for the artifacts that we've produced during our lives won't be able to radiocarbon-date them easily because items from the first half of the twentieth century will seem to have come from earlier times, and those from the second half of the twentieth century and the early decades of this one will seem to have come from later times. In fact, there won't be any objects radiocarbon-datable to our times at all, at least not correctly so. When the Suess effect takes over again after the fading out of bomb carbon, it will once again make objects seem artificially old throughout the remainder of the Anthropocene carbon tail-off. At some point, certain objects may appear to have come from our own century because their radioactivity levels seem to indicate it, but they'll merely be time-traveling imposters in museum collections of the future.

Our current time frame simply won't exist in literal readings of the geologic record. From a future historian's perspective, the tale of our present world, in a radiometric sense, will be a missing chapter that was torn completely away from the book of history.

6

Oceans of Acid

When beholding the tranquil beauty and brilliancy
of the ocean's skin, one forgets the tiger heart
that pants beneath it; and would not willingly
remember that this velvet paw but conceals a
remorseless fang.

—*Herman Melville,* Moby-Dick

Like most people, I have tended to dwell on weather, ice sheets, and sea levels when I've thought about global warming. But there's yet another aspect of the issue that has gotten less attention in the public realm, one that involves chemical changes that are far more damaging to living things than carbon isotope imbalances. Ocean acidification is one of the most ecologically important aspects of today's carbon crisis, but it is rarely discussed while we fret about things that are more plainly open to view. The climatic shifts caused by our artificial greenhouse gas buildups may arguably provide benefits in some situations as well as harm in others, and in the end they are reversible over time. But there is no obvious bright side to the irreversible, acid-driven extinction of species. As a direct threat to the aquatic life that swims, crawls, or sits beneath our planet's mostly blue surface, ocean acidification decisively tips the scales of ethical judgment against allowing our fossil fuel emissions to continue unabated.

How can oceans acidify? They seem too huge to be harmed by something as seemingly benign as air. They are also shielded from most major chemical disturbances by a defensive line of substances known to chemists as the "carbonate-bicarbonate buffering system." But although the buffering system offers some protection, it does so only up to a point. Under a sustained pollutant assault, the chemical fortress of the sea can indeed be overwhelmed.

Here's the problem: CO_2 dissolves in water. That's no great surprise if we remember that fish breathe waterborne oxygen through their gills and about a quarter of the excess CO_2 that we emit each year diffuses into the oceans. From a purely climate-centered perspective the marine uptake of our excess CO_2 resembles a helpful alliance, as if the oceans were on our side in the struggle against global warming. After all, this is how most of our carbon emissions will eventually be scrubbed from the atmosphere. But CO_2 doesn't just disappear when it enters a water body like the ocean. It morphs into carbonic acid.

The chemistry involved in this process is rather involved, but the most important concepts to grasp in this context are fairly simple. All of the main molecular characters that you need to know about in this context include carbon in their physical structures as well as in their names, but some are predators while others are prey. The top predator is carbonic acid, and its primary prey are base molecules called "carbonates," particularly the mineralized forms of calcium carbonate. Spill battery acid onto a patch of concrete pavement—cement is often refined from marine carbonate deposits—and the acid molecules will savage their target in a sizzling burst of foam that leaves behind a sullen puddle of neutralized waste salts. That kind of reaction saves lives when it drives the frothing spray of a fire extinguisher, but it snuffs them out when it attacks the shells of living sea creatures.

Carbonic acid releases tiny, positively charged hydrogen ions into the surrounding water. Their positive charge reflects the temporary loss of a negative electron from each hydrogen atom, a natural result of dissolving in water. If this concept is new to you, then it might help to think of molecular bathers whose swimsuits have been removed: imagine the naked particles drifting about in a state of nervous eagerness to replace their protective electron coverings. Any molecular passerby wearing potentially removable electrons, as do carbonates and their bicarbonate cousins, is at risk of a destructive chemical stripping.

The more such energetic hydrogen ions there are floating around in solution, the more acidic the solution is and the more corrosive it is to materials such as seashells and limestone. Keeping such carbonate-rich materials in a solid state rather than dissolving in acidified water is a bit like trying to stack hay in a hurricane.

Nobody expects the oceans to become fully acidic in a vinegary sense; it would take an unrealistically large and rapid CO_2 release to do that. Rather, they're shuffling closer to the acid-base border but staying on the basic side of the line. Chemists mark that border with a neutral pH value of 7, thereby segregating higher pH values into the realm of basic conditions and lower values into the realm of true acids (pH, or *potentia hydrogenii*, is the formal measure of hydrogen-ion concentrations). Marine scientists have watched with increasing concern as mean ocean pH has dropped during the last century. It might not sound like much, but because the numbers are drawn on a logarithmic scale, the recent pH drop of a tenth of a unit means that average marine acidity has already increased by about one-quarter. If the pH eventually falls by 0.3 to 0.4 units, as most scientists predict it will by this century's end, it will represent more than a doubling of acidity.

The most immediate victims of ocean acidification are likely to be organisms that build shells or other body structures with acid-soluble carbonate minerals like calcite and aragonite. Aragonite is the more soluble of the two, and species that rely upon it are at greatest risk. According to one source, much of the Southern Ocean's surface will be acidic enough to destroy aragonite by 2030 AD, and shallow waters in the Arctic Ocean could reach that point even earlier. If we follow an extreme-emissions path, the polar seas will be aragonite-corrosive from top to bottom by 2100 AD and will stay that way for thousands of years.

What kinds of organisms are at risk? A quick rundown of widely known sea creatures produces an inventory that many of us would be sorry to lose. Alaska king crabs and California abalones. Caribbean conch snails and spiny lobsters. Chesapeake Bay oysters and blue crabs. Classic Maine steamer clams and scallops. Irish cockles and mussels. Countless forms of starfish, sand dollars, sea urchins, and barnacles, and even the beautiful limestone edifices of coral reefs.

It can be difficult to imagine how a creature that lacks hands or even much of a brain could craft a delicately sculpted shell from much the same stuff that Michelangelo chipped and smoothed to produce his marble masterpieces. Some aspects of the process remain mysterious even to experts, but the fundamentals are fairly straightforward. Seawater carries many dissolved substances: salt, for instance. It also contains calcium atoms that have been eroded from rocks and soils on land, as well as carbonate and bicarbonate molecules that are likewise released into watery solutions by mineral weathering. Marine calcifiers gather calciums and carbonates from the surrounding liquid and stick them together to build crystal lattices of calcite and aragonite. Or, more properly, specialized cells in their body tissues do it, bit by molecular bit. Slowly, the composite molecules of calcium carbonate accumulate on the thin outer lip of a shell or the rim of a polyp's home cup on a branch of coral, spreading outward in orderly rows like raindrop rings dilating on the surface of a puddle.

Marine calcifiers dedicate much of their food energy to secreting and maintaining their shells. If the pH of the seawater around them falls closer to the acidic range, it becomes more and more difficult to coax carbonates out of solution. In addition, it requires more energy to repair the

Thais snails and their barnacle prey exposed at low tide on the coast of Maine.
Curt Stager

corrosion that results when hordes of hydrogen ions raid the carbonate storehouses of a shell or coral formation. At some point on the acidity spectrum, more calcium carbonate pairs break apart than are built, and as more and more hydrogen ions attack them, shells begin to crumble brick by molecular brick.

We don't yet know exactly which organisms can adapt to acidification and which cannot. We don't even know the full life histories of most marine animals yet, much less their potential responses to freakish chemical changes in seawater. But we do know that, at least in theory, any acidity increase could make it more difficult for a shell-producer to lure carbonate into solid form and keep it there. And in the Darwinian struggle for existence, the extra physiological investments required to overcome that chronic erosion might mean the difference between breeding and dormancy, growth and stasis, or even life and death. In the worst case, a hapless creature's hard parts might simply melt away altogether.

A seminal overview published in 2005 by Britain's Royal Society has shown that it is already too late to avoid some of these anticipated changes. Because of delays in the transfer of dissolved gas into the deep ocean, it will take many centuries for marine chemistry to return to normal even if we could halt all of our emissions right now. David Archer has estimated that the recovery process could take as little as 2,000 years or as many as 10,000, depending on whether we follow a moderate- or an extreme-emissions scenario and on how rapidly the neutralizing effects of geological weathering can restore the oceans to their present chemical balance. If atmospheric CO_2 concentrations eventually reach 500 to 550 ppm, aragonite will dissolve in the cold circumpolar seas. If CO_2 concentrations triple, then calcite will dissolve there, too. And an all-out 5,000-Gton emission could destroy both aragonitic and calcitic shells even in tropical waters.

Why are chilly places so much more vulnerable than the rest? It's because cold water can hold more dissolved gas than warm water does, so frigid high-latitude seas retain more CO_2 in solution than those at lower latitudes. Because of this temperature effect, aragonite corrosion will strike the polar regions first and hardest during winter, their coldest season.

Though most media sources still ignore this issue, planktonic lifeforms that are barely visible to the naked eye have already become the poster children of ocean acidification alerts among marine scientists. Considering the value of beauty in spreading a message effectively, it helps that some of these, most notably the micromollusks known as pteropods (TERR-o-pods), are so lovely that some biologists call them "sea butterflies." Although they share a phylum with snails, slugs, and clams, they don't creep or burrow as most of their relatives do. They swim or, more properly, fly under water. Much of their soft tissue consists of wispy, parachutelike sails or even paired wings, as the English translation of their scientific name attests; pteropod means "winged feet."

Under a microscope, many pteropods resemble translucent baby snails swaddled in billowing sheets of cellophane. In fact, real snail larvae do look a lot like pteropods, but they become clunky bottom-dwellers when they grow up. Pteropods, in contrast, seem to stay forever young. They drift about in the sunlit upper layers of the sea in huge numbers, feeding on even tinier prey and soaking up dissolved minerals with which to build fragile shells of soluble aragonite. Plankton sampling nets the world over are now hauling in pteropods whose shells show disturbing signs of carbonic acid etching and pitting, and lab studies seem to confirm the cause. To a trained eye, this is a warning flag.

Pteropods are well worth caring about in their own right, but they are ecologically important, too. In the icy waters surrounding Antarctica, a single bucket-haul can contain hundreds of thousands of them, and they support food chains that link them to penguins, seals, and whales. In Antarctica's Ross Sea, microscopic pteropods can outweigh the immense swarms of shrimplike krill. Fortunately, some two-winged pteropods—called "sea angels"—lack shells and may therefore be immune to acid corrosion. Although they have no shell, they aren't necessarily defenseless. At least one Antarctic species produces a toxin so potent that other planktonic animals occasionally seize a guardian "angel" and carry it around like a shield to protect them from larger predators. Acidification might simply favor angel-type pteropods over the shelled butterfly types, but that's not necessarily good news in terms of marine ecology. Replacing shelled species with toxic ones could be disastrous for other creatures that rely on pteropods for food.

Meanwhile, another tiny calcifier may be in trouble, as well. Satellite images occasionally show milky clouds staining hundreds of square miles of ocean off the coast of Alaska. These are single-celled algae that, at certain stages of their life cycles, look like microscopic bowling balls plastered with cream-colored hubcaps. Their long scientific name, cocco-lithophore, refers to the spherical form (*coccus*) of the cells and the miner-alized (*litho*) calcite armor plates that they carry (*phoros*) around with them.

The cumulative mass of calcite discs floating in a single Alaskan coastal bloom can weigh more than a million tons. Like tiny plants, cocco-lithophores are photosynthetic and they represent a vital base of marine food chains in regions such as these. But although their platelike armor is made of calcite rather than the more vulnerable aragonite favored by ptero-pods, some are already beginning to dissolve.

The Royal Society calculates that most polar ocean waters will soon be undersaturated with respect to aragonite; that is, aragonite will dissolve there if not maintained, at high physiological costs, by the organisms that secrete it. Undersaturation is then expected to spread to warmer latitudes, slowly inching toward the tropics from both poles. But long before the sparkling waves of Jamaica and Tahiti turn deadly to many of their resi-dents, other processes will work less visible mischief in the deepest, darkest recesses of the marine world.

Water contracts as it cools, so cold water is denser and heavier than warm water. Because of this, the long winter nights and record-breaking cold of Antarctica help to produce some of the world's densest seawater. The freezing of briny water at the surface also leaves residual salt in solu-tion, which further increases density. Gravity takes over and pulls the heavy, near-frozen liquid down to the bottoms of the southern ocean ba-sins, where it creeps like a sluggish submarine flood that can be hundreds of feet thick on the abyssal plains and even thicker in the deeper trenches and troughs. So much Antarctic bottom water forms each year that it blan-kets more than half the seafloor beneath the Atlantic and Pacific oceans, stopping its northward crawl only after Arctic bottom water overruns it; sometimes the contact zone lies as far north as the Grand Banks of Maine and eastern Canada.

Normally, this is a good thing. It's too dark down there for plants or algae to photosynthesize, so deep-living marine animals would suffo-cate without that flow. A flowerlike sea anemone need only sit still down there as slow-moving bottom currents deliver food particles to it. And as those cold, watery breezes waft gently through the animal's delicate tentacles, they also represent its sole source of oxygen soaked from the churning polar winds topside.

Unfortunately for marine life in the Age of Humans, the other gases that tumble about in those polar winds also join oxygen in the deep-sea voyage. And that includes carbon dioxide.

For many of us who think of greenhouse gases only in terms of climatic change, the polar downwelling zones may seem to be helpful storm sewers through which our CO_2 problems can drain down and out of the air. But in this case, out of sight and out of mind does not mean out of existence. Every CO_2 molecule that is pulled under in this manner represents one more carbonic acid molecule added to the ocean. And if it enters near the poles, chances are good that its acidity will sooner or later contaminate the deep-sea buffet line, a cold, dark realm that is unlike anything most of us have ever seen. Though the foundations of the oceans cover most of this planet, they are still largely unexplored by marine biologists. With little or nothing to tether our imaginations to, we tend to act as if such places don't even exist. But they do.

Despite the chill, the intense pressures, and the darkness of the deep, it is home to many species. Most of its denizens live on or within soft organic oozes, feeding on a rain of dead plankton and the occasional whale carcass. And the deep-sea bestiary is a bit light in the lovably lovely department, favoring instead the utterly fascinating. Octopi with arm suckers that glow in the dark. Vampire squids whose very name speaks volumes. Toothy, widemouthed female angler fish whose puny mates fuse permanently to their posteriors, dangling there like vestigial appendages. And in frigid waters off the coast of Antarctica, pale yellow, spindly-legged "sea spiders" the size of dinner plates that step daintily among the hulks of enormous lumpy sponges.

Population numbers below a few thousand feet depth tend to be low because there is generally not enough food down there to support much living biomass. Even so, the biodiversity can be staggering. One stretch of soft sediment two miles down and several miles out from New Jersey and Delaware yielded nearly 800 species captured in narrow vertical core tubes. And with the help of submersibles, biologists have recently found an amazing wealth of colorful sponges, crabs, and fish along the walls of mile-deep canyons in the Bering Sea. Nets and dredges have pulled up so many new species in recent years that some scientists postulate a global deep-sea tally of more than 10 million.

Most relevant to this story are the many creatures that don't swim at all and whose hard parts contain carbonate minerals. Cameras attached to remotely operated vehicles relay images of feathery red crinoids that resemble flowers but whose closest familiar relatives are starfish. Bristly sea cucumbers dot the bottom like humpbacked porcupines. Tusk-shaped scaphopod snails wriggle through the muck in search of detritus to swallow, and deepwater cuspidarid clams break ranks with their filter-feeding, chowder-stocking kin in order to hunt fleshy prey. Hermit crabs scuttle about, lugging their snail shell homes with them. And perhaps most surprising of all, there are corals, lots of corals.

Common wisdom holds that coral reefs exist only in warm tropical shallows, but in fact, two-thirds of all known coral species live in deep, cold habitats, far outnumbering the ones found in all the famous near-surface reefs of the Indo-Pacific and Caribbean. Cold-loving corals can be surprisingly abundant in waters ranging in depth from several hundred feet to more than two miles (3 km). Among the species turning up in dredges and submersible photos are precious black corals, gorgonian sea fans, ivory tree and red "bubblegum" corals, stony scleractinian corals, and twiglike bamboo corals. For some reason yet to be determined, large deepwater reefs are more common in the North Atlantic than in the North Pacific where the corals tend to be more solitary, forming diverse but diffuse aggregations scattered across the seafloor. In contrast, many Atlantic species fuse into mounds hundreds of feet tall, piling high atop the dead rubble left by their predecessors, and some form reefs whose dimensions are measured in miles. As with shallow reefs, these corals provide shelter and sustenance for a riotous wealth of other animals. One recent survey found more than 1,300 species living in, on, and around deep-dwelling thickets of *Lophelia pertusa*, the dominant reef builder of the northern Atlantic. This hidden trove of animal species rivals that of Australia's Great Barrier Reef.

The diversity and abundance of deepwater corals reflect the amount of planktonic food particles raining down on them and the nature of the waters bathing them. Under the right conditions, cells in the thin outer skins of individual polyps gradually build cuplike supporting structures of acid-sensitive aragonite. But their stony communal formations grow very slowly when food is scarce and the water is cold; some that were recently found in the lightless depths of the mid-Atlantic took thousands of years to reach their present size. For a colony that barely manages to keep its carbonate balance in the black, so to speak, it's easy to deduce what acidification could mean.

The Royal Society suggests that a rapid CO_2 rise to 550 ppm by 2100 AD might halve the annual rate of aragonite calcification in tropical near-surface reefs. A team headed by Joan Kleypas of the National Center for Atmospheric Research in Colorado, more conservatively estimates that tropical coral growth could decline by a third or more under those conditions. But however these numbers actually play out, carbonic acid buildups will be especially troublesome for deep-sea corals because cold downwelling currents shuttle huge loads of CO_2 to the bottom. Climate scientists Ken Caldeira and Michael Wickett estimate that most of the world's deepest waters will continue to dissolve aragonite and calcite even if atmospheric CO_2 concentrations level off at a mere 450 ppm, well below the peak values expected in our moderate-emissions scenario.

Truth be told, we don't know what acidification will do to most marine organisms during the next millennia of the Anthropocene. We've barely begun to discover and describe them, much less subject many of them to laboratory testing. But we already know enough to worry. According to the Royal Society, "ocean acidification will threaten the existence of cold-water corals before we have even started to understand and appreciate their biological richness and importance for the marine ecosystem." The nets of deep-sea trawlers now smash blindly through dense, brittle tangles of ancient coral in pursuit of equally threatened deepwater fish such as orange roughy, but even in more rugged and inaccessible submarine terrain and even in an enlightened future that might prohibit such destructive practices, there will be no such refuge from acidification.

On the other hand, some investigations into this subject sound more hopeful notes. A team of British biologists headed by Helen Findlay of the Plymouth Marine Laboratory recently found that certain bottom-dwelling limpets, snails, mussels, and barnacles somehow manage to *increase* the thickness of their shells under acidified conditions. Lab studies have shown that some species of plankton-dwelling coccolithophore algae can do this, too, and one North Atlantic sediment core analysis has even revealed a surprising trend toward heavier coccolith shells at the study site during the last century of increasing acidity.

When Israeli biologists Maoz Fine and Dan Tchernov exposed Mediterranean corals to high-CO_2 conditions, their subjects survived the increased acidity by doing away with their hard parts altogether. After a year of exposure to pH values as low as 7.3 (nearly ten times as acidic as today's oceanic average, and very close to the acid-base boundary of pH 7), the naked polyps were not only still alive, they were thriving. They resembled bunches of soft, flowery little sea anemones more than typical solid coral formations, but they had three times the body mass of normal polyps, perhaps because they didn't have to expend energy on depositing new limestone day after day. This hitherto undocumented transformation may explain how some kinds of corals survived extreme natural acidification events in the distant past, such as the PETM super-greenhouse 55 million years ago.

Unfortunately, even these sparks of hope glow dimly. It's still not clear how most marine creatures will really respond to future acidification or how many corals can disrobe at will. And even if many corals can go naked indefinitely, a cessation of reef-building could leave other specialized reef dwellers homeless, with unpleasant consequences for many of our descendants, as well. Coral reef fish now represent as much as a quarter of Asia's annual marine catch and feed close to a billion people.

Although some corals apparently have adaptive defenses against acidification, they'll still have to face other problems during the next centuries of the Anthropocene as well. Rising temperatures during the last fifty years have already pushed many tropical reefs close to their tolerance limits. Shallowwater corals are also picky about the depths they grow in, so they will have to keep pace with deepening seas in addition to the physiological drag of acidified water. Even the simple act of respiration could be affected. Warm waters hold less dissolved oxygen than cool waters, and absorbing that life-giving gas from the warmer, increasingly CO_2-rich oceans could become more and more like recycling stale breath from a plastic bag.

Ironically, warming seems to be benefiting some species that live on the outer fringes of their ranges, at least for the time being. New thickets of staghorn coral were recently discovered off Fort Lauderdale, Florida, where they had been absent in previous decades, and elkhorn coral is colonizing new sites in the northern Gulf of Mexico. Some marine biologists suspect that this is due to a temperature-driven expansion of suitable habitats poleward, which makes intuitive sense; as formerly cool waters heat up, they could open new territories for coral settlement just as they did during past interglacial warmings. Unfortunately, ocean acidification runs roughshod over that optimistic scenario. Even if reefs start to grow where newly raised temperatures invite them to follow, the changing chemistry of the water may still work against them.

The gradual acidification of the oceans will last a long time on the scale of a human lifespan, but like greenhouse gas concentrations and global temperatures the acidity will eventually reach a maximum value and then it will begin to decline. Mineral weathering will wash bicarbonate into the oceans and thereby help to neutralize their acid loads while also removing more CO_2 from the atmosphere. Most experts estimate that the worst phase of acid contamination could span several centuries in a moderate-emissions scenario and several thousand years in an extreme scenario.

On the other hand, any species extinctions that accompany marine acidification will be permanent. Evolution can eventually regenerate the purely numerical tally of biodiversity, but it can't re-create the unique combinations of features and interactions of species that have been lost. Too much time, too much complexity, and too much chance are behind the origins of today's organisms for us to win back what we may lose in the near future. Even if the deep future eventually does raise Earth's species counts close to the already-depleted ranks of today, it will only do so on time scales that dwarf the span of the Anthropocene. The long tail of the CO_2 curve is measured in thousands of years, but the ages of many animal and plant taxa are measured in millions.

Past warm events offer some guidance in this matter. The Eemian interglacial was not accompanied by major increases in acid-producing CO_2 concentrations, but the PETM hothouse left a permanent chemical burn on the sedimentary records of the deep sea. About half of all bottom-dwelling foram species died out during that acid bath; this alone demonstrates that the risk of extinctions is real. Coral reefs declined rapidly in the tropics during that event but they persisted in midlatitudes until large and encrusting species such as oysters, bryozoans, and red algae replaced most of them as the dominant reef builders. Some of this reorganization of marine communities may have been due to high temperatures as well as to chemical changes; most of the surviving reef corals apparently clustered in cooler, higher latitudes despite their probable higher acidity. But in any case, sizable coral reefs only returned to the oceans millions of years later, after the CO_2-rich Eocene climate optimum had passed and the world switched into long-term cooling mode.

But the PETM example also offers some hope for the future; that early ecological hit list was selective. Extinctions certainly occurred in the deep sea, but many mollusks, corals, and other calcifiers managed to survive, particularly in shallow waters. Cores from the cool Southern Ocean, where acidification would have struck early, suggest that little net change occurred among calcareous plankton species; some forms vanished but others appeared or persisted throughout the acid pulse. Few of their fossils show signs of altered shell thickness or weight despite clear geological evidence of carbonate undersaturation in their surroundings.

At this point, we can only speculate about which species will disappear because of future acid buildups, and we can only wonder how their demise will affect others around them. But we can be sure that important ecological changes will occur, and that at least some will be unwelcome. Oceans that lose selected calcifiers from the complex webs of marine life will be like a hockey team that suddenly loses a random assortment of members to flu shortly before the playoffs. The replacement skaters, if they even show up, may be poorly matched to the long-termers, and the patchwork team's performance could be quite different from what they and their supporters had hoped for.

Consider starfish and sea urchins. They build their protective spines and shells from calcite, and their larvae secrete a particularly soluble high-magnesium form of the mineral. Starfish eat barnacles and mussels; urchins eat seaweeds and other algae. Together, these hunters and grazers clear away small patches of living space on crowded stony substrates where other bottom-hugging settlers can later attach; cage off a section of seafloor to exclude urchins, for instance, and the enclosures sprout lush algal gardens. Lab experiments show that the larvae of some stars and urchins stop growing and develop dangerously brittle shells when CO_2 concentrations are doubled. Loss of these keystone species to acidification could radically alter the community ecology of seaweeds, shellfish, and barnacles.

In a paper that appeared in *Nature* in 2008, a team of scientists led by British marine biologist Jason Hall-Spencer described a natural demonstration of what carbonic acidification has already done to such creatures. They studied a volcanically active region of the Mediterranean seafloor near Italy's east coast where CO_2 emerges from cold gas vents and acidifies the local waters by as much as half a pH unit, which is close to what is soon expected to happen at high latitudes. They were careful to select vents that don't release other toxic substances such as sulfur, so the example may offer a reliable hint as to what such changes could do on a larger scale in the future.

Corals were common in the general vicinity of the vents but not anywhere close to them, and the encrusting coralline algae that often help to hold reefs together were also absent from the acidified zones. Sea urchins and many snails were either killed or driven out, too. On the other hand, most organisms that lacked soluble hard parts did well. Many-tentacled sea anemones crowded like soft bouquets near the vents, and CO_2-loving seaweeds also did well, perhaps because of the dissolved gas, the lack of crowding from the missing bottom-dwellers, and the absence of grazing urchins. It all sounds rather complicated, as one might expect from a diverse marine environment such as this, but the main point is that local, natural carbonic acidification has clearly changed what lives on that patch of seafloor.

Much larger habitats and economies may also be at risk where cool, increasingly acidic water wells up from deeper layers of the sea to the surface. Such upwelling zones are among the world's most productive habitats because they deliver nutrients into sunlit shallows where algae can feed on them and grow like grass on a fertilized lawn. Tiny planktonic animals graze on those algae, and they in turn support great shoals of fish such as sardines and anchovies. Sea lions, squid, dolphins, and other predators feast in such places, and local trawler nets bulge with valuable fish protein. A quarter of the global fish catch comes from near-shore upwellings along the western coasts of South America, California, Spain, southern Africa, and New Zealand.

Oceanographers are already reporting a drop in pH values in California's upwelling zones from excess CO_2 that entered the deep ocean decades ago, and these sorts of places will bring early signs of acidification to the tropics before the rest of the low-latitude shallows catch up. Predicting what components of these tightly linked food chains might be lost to acidity is merely a guessing game at this point, but sporadic natural events in Peru and South Africa hint at how bad things might become if even one link fails.

Normally, the upwelling of cold currents along those coasts makes the algae so abundant that it turns the surf foam green, but sometimes that upward flow slows down or even stops. In Peru that change is caused by El Niño climate disturbances and in southwestern Africa it is caused by weather disruptions associated with the coast-hugging Benguela current, but the results are similar in both cases; the algae wane, then so do the fish and the things that eat them. An unusually strong El Niño in 1983 temporarily destroyed the Peruvian anchoveta fishery, and Benguela events can decimate catches, as well. The algae that usually live in these habitats lack carbonate shells so they might not suffer much from a drop in pH, but early acidification there might nonetheless harm shellfish or crabs that live along those coasts. We simply don't know yet.

Living ecosystems can be surprisingly complex, of course, and treating one of them like a simple chemical equation in which A minus B produces C can be misleading. Hopefully, we'll know a lot more about the ecological effects of acidity on oceans sooner than later. Since the Royal Society's report was published, many other scientific organizations ranging from the U.S. National Science Foundation to the International Geosphere Biosphere Program have launched research initiatives on ocean acidification.

But time might be running out even faster than we think. Ambitious geoengineering schemes designed to turn sunlight away from Earth with high-altitude reflectors might be able to cool climates down below, but even if they succeed they will still leave the excess CO_2 in circulation. Worse still, plans are also afoot to pump massive amounts of atmospheric

CO_2 into the deep oceans. For those who focus exclusively on ways to avoid temperature changes, that solution seems to be quite sensible. The sea can absorb most of what we emit, so why not help things along with aquatic carbon sequestration?

To those with a broader environmental perspective, however, this can seem like trying to douse a fire by pouring gasoline on it. As Ken Caldeira and Mike Wickett noted in a recent paper in the *Journal of Geophysical Research,* injecting CO_2 into seawater amounts to trading "greater chemical impact on the deep ocean as the price for having less impact on the surface ocean and climate."

Only time will tell us if a general outcry over ocean sequestration stops it in its tracks. But relatively few outside the scientific community are discussing the dark side of this issue yet, much less the deadly effects of carbon emissions on marine ecology. One of the most influential articles about acidification in the popular press was Elizabeth Kolbert's masterfully written piece that appeared in *The New Yorker* in 2006, titled "The Darkening Sea." We need more of that, and soon. But to turn the present situation around in time to prevent widespread damage will require rallying public sympathy for squishy little sea creatures with unpronounceable names that most of us don't even know exist. Could T-shirts and banners emblazoned with "Save the coccolithophores" or "I love *Lophelia*" raise much of a response outside of narrow scientific circles?

Ocean acidification represents one of the most compelling reasons to control our emissions, not only for ourselves but for the sake of countless other species that share this water-dominated planet with us. Temperatures and sea levels will also change in response to our carbon emissions pulse but they will eventually recover in the deep future. When species die out, however, they don't return. Extinction, far more than global warming, is truly forever.

7

The Rising Tide

*And the waters prevailed exceedingly upon the earth;
and all the high hills that were under the high
    heaven were covered.*   —Genesis 7:19.

*And when the seventh day came, the flood subsided
from its slaughter, like hair drawn slowly back
    from a tormented face.*   —Epic of Gilgamesh

Invisible but important chemical disturbances in the air and oceans attest to the reality of global carbon pollution, and they also warn of more changes to come. One of the most troubling of those changes, even for people who care little about carbon chemistry or unfamiliar aquatic species, is also invisible but only in the sense that it proceeds too slowly for us to watch it unfold on a daily basis. As the world warms, the physical shape of the sea is changing.

Global sea-level rise is, in a psychological sense, a mirror image of marine acidification. The latter process is underreported and it rarely generates strong emotional responses among the general public. But sea-level rise is quite another matter. One can easily imagine its effects on coastal settlements, and many of us grew up on tales of epic floods in the days of Noah or Atlantis, so the primal fear of advancing waters runs deep. As a result, ocean encroachment is widely recognized as a problem, but the details of it are often misunderstood. What, exactly, *is* sea level? How high can it climb, and how fast? How dangerous is it? And what will happen to the oceans far out on the long, cooling tail of the CO_2 curve?

Defining and measuring sea level itself sounds straightforward enough, doesn't it? "Sea level" is simply the height at which the surface of the ocean rests. Just measure it all over the world and calculate the average. But think more carefully about actually having to go out and do it, and you'll understand why I have been struggling to nail down various aspects of the concept since tenth grade.

My first encounters with this surprisingly complex subject began in Mr. Alibrio's high school physics class in Manchester, Connecticut, during the 1970s. One day he abruptly called my name from the front of the room, brushing with one hand a military-style crew cut that was bleached by age. "Stager! What is sea level? Look it up and tell us next week."

A week later, I rose to say that sea level was an average of many measurements of sea-surface heights. "Not good enough!" Alibrio roared. "I want to know exactly *how* you measure it, too."

After another week, I reported that sea level isn't measured with rulers but with barometers. A former surveyor had told me that the elevations of mountaintops are calculated relative to atmospheric pressure at sea level. "Still not good enough!" came the verdict. "Think about it: air pressure goes up and down from day to day as the weather changes. How can you get a standardized value for sea level with a wavering tool like that?"

And thus it went. The surface of the sea is roughened by waves, not to mention rising and falling tides, so what part of it are you supposed to measure, and when? And what is it measured relative *to*? If you measure it relative to a point on land, then you have to account for the long-term shifting of seemingly solid ground as well as of the more obviously fluid oceans. Many regions of Earth's crust are surprisingly mobile, drifting and bobbing like gigantic rafts afloat on the magma-filled mantle. Most major movements of the Earth are slow, the result of processes such as tectonic plate collisions or local groundwater removal, but they can still be significant. If we were to measure sea level relative to such places, then we might falsely conclude that it is rising or falling faster than it really is.

Because of this often unsteady relation between land and water surfaces, the elevation of Britain's official brass sea-level benchmark is represented by measurements at a single location, Newlyn. The data were compiled by observing the position of the local sea surface on a tidal staff every fifteen minutes for six years (yes, that means day and night all year long) between 1915 and 1921, and all other elevations in England, Scotland, and Wales were originally based upon that standard point on the national map. Other countries use different reference points, and even within a single nation the varying needs of different user groups can lead them to use different standards. For instance, nautical maps typically emphasize the hazards of submerged rocks and sandbars by measuring water depths against a chart datum below which tides rarely fall, while warning signs on bridges and low-hanging cables present their heights relative to mean high water.

I never did give Mr. Alibrio a satisfactory answer to his question, but that may well have been the real point of the lesson. As the years passed and I taught science classes of my own, I began to realize that my old teacher wasn't so much trying to torment me as to introduce me and my peers to the remarkable depth and detail behind such seemingly simple questions.

Nowadays, state-of-the-art satellite systems use radar and other energy emissions to monitor and calculate surface elevations worldwide. With such tools we can even see where large sweeps of saltwater bulge or dip by as much as several feet because of tides, currents, and the gravity fields emanating from submarine mountains.

But how do satellites calibrate their measurements as their orbital paths slowly decay and move closer to the ground? Do they drop sounding lines to Earth to check the accuracy of their measurements? From my work on deep lakes, I know that supposedly well-calibrated electronic gadgets can give readings quite different from those that you get by lowering a heavy lead weight on a graduated rope.

And what, exactly, does a high-flying spacecraft measure sea level relative *to*? Does it record how long it takes a radar signal to bounce off the ocean and return, and then calculate how many miles above the sea surface it is at any given moment? If the satellite measures a slightly shorter distance the next time it flies over that spot, does it mean that sea level rose or that slight losses of energy lowered the long flight path a bit, or was it some combination of both?

Today, multiple satellites and global positioning systems give us a diverse selection of methods for determining surface heights, using three-dimensional, space-based networks of interlocking angles and reference points for drawing maps, calculating locations, and measuring sea levels. But even the most sophisticated satellites still don't help us with truly long-term trends because their records cover only the last few decades. Instead, we must turn to geohistorians if we want to know what major sea-level changes have been like in earlier times.

Composite charts of sea-level positions since the peak of the last ice age combine readings from fossil deposits in places such as Tahiti, Barbados, western Australia, and the Red Sea. Those charts show that sea-level change itself is nothing new. About 20,000 years ago, when continental ice sheets held the most water away from the oceans in frozen form, you could have walked from what is now Cape Fear on the sandy coast of North Carolina all the way out to the edge of the continental shelf, roughly 50 miles (80 km) to the east. The ancestors of Native Americans walked from Siberia to Alaska on a land bridge that now lies beneath the storm-whipped Bering Strait, and Paleo-Europeans could have wandered across what is now the English Channel without even seeing a beach or sizable body of water. The last ice age dropped ocean surfaces by about 400 feet (120 m) for tens of thousands of years, and similarly severe ice ages struck repeatedly during the last 2 to 3 million years. If satellite cameras could have filmed, say, the Carolina coast on a continuous basis over that time period, a fast-forward run of that footage would resemble a home video of your favorite beach, with wave after wave alternately swallowing and disgorging the border between land and water.

The last time we came out of a glacial deep freeze, it took several millennia of melting for global sea level to stabilize near today's elevation after eating hefty portions of beachfront. The scale of that encroachment outstrips anything in the works for us during the Anthropocene; we have only enough terrestrial ice left now to raise future sea levels by a little over half the postglacial amount. But it didn't bother our nomadic Stone Age ancestors nearly as much as smaller future rises could bother our more numerous descendants, whose dwellings and possessions won't be moved as easily as mammoth-skin tents once were.

How did that last great rise compare to today's? From 1993 to 2003, ocean levels crept upward by the thickness of three fingernails per year (a tenth of an inch, or 3.1 mm), a pace that was nearly twice as fast as the average for the full twentieth century. A French team led by geophysicist Anny Cazenave reports that the pace slowed somewhat from 2003 to 2008 (2.5 mm per year), perhaps because the global warming trend also slowed briefly on its irregular path to higher temperatures. By comparison to, say, the recent three-fingernail rate, the average postglacial recovery was about four times as fast, and several short but intense episodes sped things up even more from time to time as ice sheets surged or collapsed. For example, about 14,500 years ago a major meltwater pulse lifted sea levels by 50 to 65 feet (15 to 20 m) at more than ten times our modern rate—still much too slow to watch from shore, but those incremental changes reshaped coastlines dramatically over the course of centuries.

As in the past, today's rise reflects both the melting of land-based ice and the expansive effects of heating that swell seawater just as rising temperatures fatten the liquid in a thermometer tube so that it rises higher and higher the warmer it gets. As the world warms further, both expansion and polar melt will continue, but the magnitude of the rise at any given time will probably be most strongly influenced by what happens to our remaining ice.

Only three frozen masses are still large enough to change sea level dramatically, and they're sitting in Greenland, West Antarctica, and East Antarctica. Satellite-based measurements show that the first two have been losing weight since 2002, while the East Antarctic sheet may actually have gained a bit due to snowfall and exceptionally cold temperatures around the South Pole.

Seasonal melting in Greenland, which holds roughly 10 percent of the world's ice, now drives about a tenth of the current global rise. If the whole thing eventually liquifies, it could raise sea level by 23 feet (7 m).

The loosely anchored West Antarctic ice sheet is the least secure of the bunch. Covering nearly 800,000 square miles (2 million km²), it contains enough meltable water to hoist sea levels by 16 feet (5 m) or so. Winter temperatures on its narrow western peninsula are warming faster than in any other place on Earth, up by 10 to 11°F (6°C) since 1950, and just how unstable it is remains unclear. But for now it contributes only slightly more than Greenland does to sea-level rise.

About 80 percent of the world's remaining ice—the gigantic East Antarctic ice sheet—seems to be holding steady for now because its surface is so high and cold that relatively little melting occurs there at present. Some simulations of a wetter, warmer world show more snow falling there in coming centuries, which might actually thicken the vast interior as the lower, warmer coastal margins diminish, and it might therefore slow global sea-level rise somewhat, at least at first.

On the other hand, it's hard to imagine much ice surviving a long-term greenhouse future. Some experts foresee Greenland's ice cap destabilizing if temperatures rise more than 2 to 7°F (1 to 4°C), close to the range of temperatures expected for a moderate-emissions scenario. And most who have baked the more persistent south polar cap to death in computer simulations expect it to destabilize if temperatures rise by 9°F (5°C), which would be likely in an extreme-emissions scenario. But even more of a threat to polar ice than the magnitude of future heating will be its immense duration, at least 50,000 to 100,000 years, with temperatures at or above those of today.

How might the world look without any ice at all? For a glimpse of that possible future, we can turn to computer programs that overprint 230 feet (70 m) of virtual seawater onto topographic charts of the modern world. On such maps, the southeastern United States looks like it's been chewed by sharks, with the entire thumb of Florida bitten off. The sharks also gnaw the Yucatan Peninsula to a stump, and eastern China is mauled most of the way in to the Tibetan Plateau. Such striking images help to feed a rising tide of anxiety as people struggle to understand what human-driven sea-level change means. But how high the ocean can eventually rise is only one of the things we should keep in mind here. We also need to know how *fast* it will rise. And that's where we can be misled by short-term thinking.

Remember the last time you considered the magic of compounding interest or some amazing, and possibly true, statistic like "the average human eats a pound of dirt every year." Most of the shock value of such revelations arises from misperceptions of the time scales involved. We tend to focus on becoming millionaires while overlooking the need to put money aside consistently for many years, or we imagine digging into a bucketful of soil rather than gulping a few grains of grit along with our daily salad. Drinking several gallons of water all at once can kill us by diluting our body fluids, but drinking the same amount over a couple of months can kill us through dehydration. The key factor here is time.

When, for example, scientists warn that sea levels could surge in response to a "collapse" of the West Antarctic ice sheet, most of us envision the ice crashing down as rapidly as a building might disintegrate in an earthquake. Fevered media coverage of calving glaciers and ice shelves breaking loose from shore also magnify that impression. But geoscientists and lay folk mean different things when they use that word "collapse"; what seems fast to a glaciologist might sometimes be better described as "less sluggish than usual" to most of us. So how long would such a collapse really take?

When a noted glaciologist whom I met at a recent conference mentioned that a "catastrophic" slide-off may be imminent in western Antarctica, I pressed him for details. After some hesitation, he guessed that such a collapse would take decades at least, possibly a century or more. That time frame fits well with an article in *Science* that was published recently by geophysicist Charles Bentley, who reckoned that the right combination of warming, rising sea level and loss of supporting ice shelves could make the western slab "collapse in as little as a century by catastrophic grounding-line retreat."

The probable decades- to centuries-long duration of such an event doesn't mean that the change would be insignificant. It simply reflects the enormous quantities of relatively slow-moving ice involved; the sheet is so large that it would take years for all of it to cross the many miles from land to ocean. Recent evidence suggests that some sections of West Antarctica's interior ice are actually thickening rather than thinning, and that any major slide-off would probably be partial, leaving sizable remnants stranded on the jagged spine of the peninsula. Even so, the total marine displacement could still be on the order of 10 feet (3 m).

And what might such a sea-level rise look like from shore? Not much, if you were to sit there and try to watch it happen. Changes of similar amplitude happen all the time in Maine's picturesque Muscongus Bay, though they're compressed into brief 10-foot tidal cycles. I've happily sat there long enough to watch the slowly rising tide creep in among the cobbles and rockweed and heard the barnacles hiss as they prepared for the return of the shifting waterline. But I wouldn't want to have to wait there long enough, even amid the crying gulls and beautiful fir-clad islands, to watch a global sea-level surge swallow the shore. Not even the fastest rises of past deglaciations would have been visible to the average eye. Like the tides, such changes require time-lapse imagery or a well-trained imagination to be seen for what they really are.

Low and high tides at Muscongus Bay, Maine. *Rick Ylagan*

The last IPCC assessment report estimated that sea level could climb 1 or 2 feet (0.3 to 0.6 m) by 2100 AD, depending on how much carbon we emit. That would average out to nearly twice today's rate, though more recent estimates double or triple it. But even those faster paces wouldn't represent a frothing shoreward dash in the sense that most nonscientists imagine it. To help put this issue of rates into perspective, keep in mind that mean sea level has already risen by about 7 inches (18 cm) during the last century. Most of us didn't notice the change because it was so slow. Unfortunately, some of us may therefore doubt that it's even happening at all.

This is an important point. Sea-level rise is a serious and regrettable consequence of our carbon pollution, but not necessarily in the ways that many sources portray it. On the scale of a human lifetime, real sea-level rise won't much resemble a biblical deluge except in certain situations, most often when a local storm surge pushes floodwaters farther inland than usual and catches formerly safe inshore communities by surprise. In theory, most coast-dwelling people will be able to monitor the slow encroachment and issue fair warning to those living in the newly expanded danger zones. Whether they actually do so, of course, is another matter.

There is one kind of situation, however, in which an initially slow sea-level rise really could cause sudden and massively destructive changes on a regional scale. It could happen where a low geographic barrier holds the ocean back from a local depression, a setting that actually developed at least once in Asia Minor. What is now the Black Sea used to be a freshwater lake lying below sea level, isolated from the much larger Mediterranean by a narrow neck of land where Istanbul now stands on Turkey's Bosporus waterway. About 8,000 years ago, ocean levels were nearing the end of their steep postglacial climb when the water discovered a low point and slid a probing finger over the barrier.

Then erosion took over. Within a matter of months, a torrent hundreds of times more powerful than Niagara Falls was pouring into the Black Sea basin, raising its surface by as much as 6 inches (15 cm) per day until, two or three years later, it fused with the Mediterranean 500 feet (152 m) above the original lake level. At least that's what preliminary historical reconstructions showed; follow-up studies have suggested that the rise was more like 115 feet (35 m), quite a bit less than the original estimate but still enough to submerge a tall building.

Whatever the exact magnitude of the surge was, thousands of square miles of land disappeared within that first year. Countless houses, hearths, and bits of crockery were submerged and preserved beneath what are now a hundred or more feet of saltwater. The resulting dispersal of farming communities may have contributed to the spread of agriculture through Europe and Asia, and some historians have noted possible echoes of the disaster in a flood myth that appeared in the Babylonian *Epic of Gilgamesh* some 3,000 years later, as well as in the younger tale of Noah and his ark.

Fortunately, current coastal geography doesn't presage any new megafloods of the Black Sea type, though coastal cities that already lie below sea level are indeed vulnerable to the overtopping of artificial dikes and levees, and one or more notable ice-driven surges may still be in the cards for us or our descendants. But even gradual sea-level rise will turn the world's coasts into malleable fluid lines rather than solid boundaries, and it can radically reshape the edges of the continents over the long term.

Such changes are easily envisioned with the help of computer generated maps, several of which are freely available online; they generally focus on the initial single-meter shifts that are likely to occur within the next few centuries. The first one I encountered was created by Jeremy Weiss and John Overpeck at the University of Arizona's Department of Geosciences. The color scheme for the maps was wisely chosen; the oceans are a lovely deep blue, and the land a rich green. But the submerged zones are bright, sanguine red. Your eyes are drawn straight to them as though they were wounds on a beloved pet.

Most of those wounds don't run very deep even for a multimeter rise, and when you look at entire continents most of the problem areas make only a thin red rind along the edges. But zoom in on spots where the land slopes gently, and things become more interesting. The biggest bleeders on North America are located on the low coastal plains that stretch from the Tex-Mex border to eastern Virginia. After only a 3-foot (1-m) rise, the Florida Keys and the Everglades sink under a crimson tide and the Mississippi Delta seems to drip blood, with New Orleans swimming in the center of one great, hanging drop.

Scanning the rest of the planet, other hot spots glow like fiery beacons; San Francisco Bay, much of eastern China, the southern tip of Vietnam, Cameroon's port city of Douala, the Dutch interior, the southwestern rim of Denmark, and the broad deltas of the Nile, Niger, Orinoco, and Amazon rivers. And this is only after the first puny step of 3 feet. By geographic happenstance, additional rises of up to 20 feet (6 m) seem to add less dramatically to the redness, so many of the most notable inundations will happen early rather than later.

Inspecting the future in this manner is somewhat akin to snorkeling on a coral reef. You see something interesting, hold your breath, and dive down to take a closer look. But it's important to come up for air, too. These maps can be terrifying to behold, especially when they foretell the fate of a place that you happen to know well. But is terror the most suitable response?

I'm not exaggerating when I use the word "terror" here; sea-level rise really does scare people, and I suspect that some who use it to raise public awareness about climate change don't realize how much unhelpful panic they can cause. One set of online maps, posted by British computer program designer Alex Tingle and titled Firetree, displays comments from viewers around the world. Here's a sample of what they were saying recently:

"This page works well if you want to know if your home will be under water. Tell me if it's possible to add some time control, like year 2015, etc."

"When I built my dream home I tried to make sure that it would be above future sea levels so I won't drown in fifty years. . . . I used your work to check, and yes, I'll be fine."

"All people that live near coastal areas should move within the next twenty years . . . by 2050 sea levels will have risen 10 feet."

The first of these writers apparently expects extreme flooding to happen within less than a decade. The next person seems to believe that it will be fast enough to drown her. And the third one's impression of the rate of rise is probably off by an order of magnitude. Such comments help to illustrate how poorly most of us grasp the time scales involved. Artificially accelerated video depictions can also make things worse by printing indelible images on our memories that speak louder than words and logic.

Firetree's so-called flood maps and the bloodred colors in other, more aptly named "inundation" or "sea-level" maps aim for maximum attention, but in doing so they can overshoot the mark and strike real fear into the hearts of many who might be better served by more sober descriptions and depictions. The term "flood" itself, when used in this context, can evoke an unrealistic sense of speed and deadly destruction. Nonetheless it is used routinely, even in the scientific literature.

In typical fashion, a recent paper in *Science* tells us that the melting of Greenland could raise sea level significantly, thereby *"flooding* much of South Florida." However, a rise of 23 feet (7 m) over several centuries, which averages out to something close to today's sluggish rate, isn't what a normal person would call a flood. Such a choice of words can confuse people and intensify conflicts between concerned citizens and climate naysayers who focus on different aspects of the message. No, sea-level rise isn't going to *flood* southern Florida; only storms, tsunamis, and swollen rivers work that quickly. But the scientists aren't lying, either; the ocean really is going to submerge or *inundate* southern Florida, albeit over the course of many years.

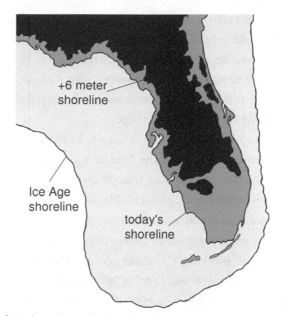

Map of Florida's shorelines during the last ice age, today, and in the future after a long-term sea-level rise that could result from a moderate 1,000 Gton emissions scenario. *After maps posted online by the University of Arizona, Department of Geosciences*

To put it another way, panic is not an appropriate response to this problem, but neither is complacency or denial. Sea-level rise will be almost imperceptibly slow for the most part, but this certainly doesn't mean that it's unimportant. We'd be better off without it, and the longer we continue to release the greenhouse gases that are causing it, the more land it will eventually consume. Over the long term, it will force maritime nations to invest again and again in the costly uprooting and rebuilding of coastal towns and port facilities, and some places—especially low-lying islands and coastal plains—will be lost for thousands of years. That's serious business, but I believe it needs no overstatement to merit our attention.

Not everybody agrees with me, of course. James Hansen, NASA's most widely quoted advocate for taking climate change seriously, is known to occasionally focus public attention on the extreme end of the list of possible futures to avoid. He has challenged the IPCC's recent estimate of a foot or two of twenty-first-century sea-level rise by proposing that "explosively rapid" ice sheet collapse could force polar melt rates to double every decade and thereby hoist the ocean's surface another 16 feet (5 m) by 2100 AD. That would involve an average rise of 2 inches (5 cm) per year, almost a twentyfold acceleration of the current rate. He urges the research community to sound a more vigorous alarm about it and suggests that any "scientific reticence" about this issue reflects political pressure to minimize concerns about climate change.

Some scientists share Hansen's views, but others do not, partly because they consider caution in the face of uncertainty to be the hallmark of good science rather than inappropriate reticence and partly because they doubt that enough unstable ice exists to launch a large deglacial-style ocean surge in modern times. A more recent review of the subject headed by glaciologist Tad Pfeffer from the Institute of Arctic and Alpine Research in Boulder, Colorado, concluded that a twenty-first-century rise of more than 6 feet (2 m) is "physically untenable," and that the most likely outcome will be a rise of roughly 2.5 feet (80 cm). But at this early stage in our understanding of ice sheet dynamics, we simply can't be sure whether, when, or how much of Greenland or Antarctica might de-ice in response to Anthropocene warming.

This is the kind of situation in which a long-term historical perspective is most helpful. Can a shrunken remnant of polar ice really launch a major sea-level pulse today? "Definitely," says Paul Blanchon, an expert on ancient reefs and sea levels based at the National University of Mexico. "Drowned coral reefs are the smoking gun that too many scientists ignore," he told me recently. Paul and his colleagues have used fossil reef deposits to reconstruct century-scale jumps in ocean levels at the end of the Eemian interglacial and shortly after the close of the last ice age. "The history of partial ice sheet collapses seems to give a tight range of rise rates of several centimeters per year," he explained. One abrupt rise during the late stages of the Eemian averaged 2 inches (5 cm) per year, an order of magnitude faster than today's rate, and lasted for about a century. "And another surge about 8,000 years ago coincided with the drowning of many Caribbean reefs. Corals eventually regrew farther upslope several centuries later."

The candidate most likely to repeat those reef-drowning events of the past is the West Antarctic ice sheet. For the sake of argument, and to illustrate the need for more information about ice and oceans, let's imagine a scenario in which a little more than half of it begins to slide off tomorrow. An associated sea-level rise of up to 10 feet (3m) would presumably occur between now and 2100 AD because the streams of ice would displace surrounding waters as soon as they hit them. But the sea-level responses to the dumping of so much ice could be more complex than one might imagine.

The West Antarctic ice sheet is so large and thick that the gravitational attraction generated by its enormous bulk makes seawater within 1,200 miles (2,000 km) of the Antarctic coast bulge slightly upward. Were all that ice to disappear, the local sea surface would slosh away toward the Northern Hemisphere, there to raise sea levels higher than average in some places while lowering them elsewhere. The slow rebounding of down-warped bedrock on the de-iced peninsula might also destabilize sea levels even further.

The upshot of all this is that the risk of relatively abrupt sea-level rise due to global warming is real, but the details of how probable it is and exactly how it might proceed remain unclear. We can be sure, however, that some sort of rise is going to happen for a very long time to come, and it will involve significant human and ecological costs.

A team of scientists at the University of Kansas have recently combined submergence maps with current population data in order to estimate some of those costs. According to their study, the first 3 feet (1 m) of sea-level rise will cover nearly 400,000 square miles (1 million km²) of coastal land and could displace more than 100 million people, nearly half of them in Southeast Asia. Northwestern Europe will lose about 13,000 square miles (34,700 km²) and might eventually see 12 million citizens uprooted. And in the southeastern United States, an expected loss of 24,000 square miles (62,000 km²) could displace more than 2.6 million people.

The study's basic premise is sound, but some of the details remain open to question. We know people will be displaced as the ocean swells upward, but we can't know their numbers exactly because we don't know how demographics will change in coming decades and centuries. Regardless of how many people are affected, how might that kind of sea-level rise affect their daily lives?

The speed of the change will be at least as important as its magnitude. If we postulate a rise of 3 feet (1 m) at twice today's rate, then it would occur over the course of nearly two centuries. Even if we shorten the time period to one century, the displacements would be spread out over several generations and thousands of miles of shoreline. If an abrupt surge of the Eemian sort should occur instead, speed those changes up by yet another factor of 5. But even that would not be like having your kids swept away from the beach while you're at the snack bar. Living with sea-level rise will be more like bay water leaking into the Manhattan subway system year after year until the pumps are finally turned off and the rail tunnels are abandoned to become submarine caves.

For most of our descendants, the coming affliction will be more of a chronic ache than acute agony. This was already recognized as early as 1975 when a select gathering of prominent geoscientists met in North Carolina's Research Triangle Park to discuss the future of greenhouse gas pollution, one of the first major meetings of its kind dedicated to the subject. In the published proceedings of the conference, the moderator concluded: "It was generally agreed that sea-level rises would be more of an expensive annoyance than a catastrophe." The relatively mild concern expressed back then was not due to some misguided belief that the coming changes would be small. If anything, the rates were overestimated; the long-term sea-level rise under discussion was expected to average several centimeters per year, much like Hansen's short-term worst-case scenario. In today's more highly charged media atmosphere, some of those same scientists now speak in apocalyptic terms, but the earlier conclusion, which was made in a more sober professional setting, still makes sense to many experts. As one of my own colleagues put it with a wry smile, "You're not going to die from sea-level rise. But if places that you care about succumb to it, then you might well *want* to die."

In the case of Europe, a climb of 3 feet (1 m) per century could force 120,000 people to migrate inland within any given year. That's a lot of uprooted people looking for lodging and employment, but housing and employers can move, too, and many Europeans already change homes or jobs for more mundane reasons without triggering widespread panic or social unrest. In the United States, 10 to 20 percent of the population already pulls up stakes every year; applying this statistic to western Europe's 400 million souls yields a ballpark figure of 40 to 80 million routine annual migrants and suggests that relatively rapid sea-level rise might add 3 to 7 percent to that total.

A focus on population numbers alone, of course, doesn't capture the full spectrum of impacts on sea-level refugees. Those who will be displaced are likely to be more closely tied economically, culturally, and emotionally to the sea than the average mainlander, and being forced to move from a sinking shoreline into an already crowded interior could mean the loss of livelihoods as well as homes.

The slow pace of the changes will allow societies to respond in more complex fashion than a sudden, uniformly unpleasant flight of refugees to higher ground. In the wealthier countries, home prices and insurance rates are linked to "location, location, location," and in oceanside nations they will be influenced by their positions on inundation maps. For some, this will be an incentive to move preemptively or to avoid oceanside properties altogether, but others might actually be lured shoreward by perennial crops of real estate bargains and short-term business opportunities there.

Consider the case of Amsterdam. In early medieval times it was an isolated, nondescript village that lay several miles inland from the ocean. Where the heavily diked Ijsselmeer harbor now stands, a broad lowland once blocked access to the North Sea. Naturally rising sea levels slowly chewed away at the Dutch shoreline until they opened a profitable sailing route to the maritime trade lanes and the rest of the world. With or without climate change, life on low coastlines always involves risks, and North Sea storms wrought terrible havoc in that region from time to time, including the Saint Lucia's flood of 1287 and the Grote Mandrenke ("Great Drowning of Men") of 1362 that killed tens of thousands of people. But without long-term sea-level rise and the new connection to the Atlantic, Amsterdam might never have become the glittering cultural and economic center that it became in later centuries.

Situations such as this will complicate the story of sea level in coming centuries, as well. As the creeping zone of inundation slowly consumes the latest array of oceanfront settlements, a migrating band of social and economic change—a "zone of anticipation," if you will—may also move inland ahead of it. As the next settlements in line realize what's coming, they will rush to prepare for the conversion of their landlocked setting into coastal status. In the honeymoon years between the ocean's arrival at a city's doorstep and the inevitable submergence, targeted towns may blossom from a resultant boost to trade and tourism. Sites with steeper slopes will have longer honeymoons than those on flats, where rising waters will consume the land more rapidly. If hardware and expertise that can be mobilized from the zone of inundation sells cheap, then outfitting the next string of port facilities could be undertaken at fire-sale prices.

But it won't be all positive, either. Rising sea levels will require constant and costly upgrading of water barriers before they are finally abandoned along with the properties they were built to protect. And accidents will happen, too. Occasional breaks in dikes have repeatedly devastated Dutch towns that lie below sea level, and in 1421 gaps in several dikes flooded seventy communities during a high storm tide and killed as many as 10,000 people. At best, the boom times in the zones of anticipation will merely represent short-term parties before a final demise.

The least prosperous nations, of course, are likely to suffer the most. Low-lying Bangladesh is already sinking under its own weight because of sediment compaction on the broad floodplains of the Ganges and Brahmaputra rivers, and its residents have long been caught between the hammer and anvil of seasonal floods and coastal storm surges. But rising sea level threatens to obliterate much of that country altogether, and the displaced people will have no easy escape route because international borders restrict their migration. Fortunately, it will take a long time, probably several centuries, for the sea to nibble its way very far inland because much of the terrain beyond the coastal floodplains slopes up toward the Himalayas. According to the University of Arizona maps, a vertical 6-foot (2-m) rise would submerge approximately one-fifth of Bangladesh, and a 20-foot (6-m) rise would consume about half of it.

Sea level will continue to climb long after the thermal maximum and climate whiplash phases pass, and where it finally levels off will depend on the height and timing of the temperature peak and how much land-based ice eventually melts into the oceans. A moderate-emissions scenario probably stops well short of a total polar meltdown, but an extreme 5,000-Gton scenario could leave the planet more or less ice-free—

and sea levels hundreds of feet higher—for thousands of years. If the ocean surface climbs 230 vertical feet (70 m) at Pfeffer's predicted rate of 2.5 feet (80 cm) per century, then we might expect sea level to continue rising for the next 9,000 years or so before leveling off.

But that's just the beginning. When the phase of net melting finally ends in the deep future, coastal peoples will have to deal with another, even longer-lived aspect of maritime change. As the air and oceans cool and snow begins to accumulate once more in the polar regions and high mountain ranges, the new challenge will be long-term sea-level retreat.

Those many millennia of future cooling will bring much slower changes than warming does because meltwater won't surge abruptly back out of the oceans. At first, most of the retreat will be due to slight contraction as the seas cool down, but continental reserves of ice will eventually be rebuilt by the gradual accumulation and compression of winter snow. Two miles (3 km) of Greenlandic ice can represent more than 100,000 years of growth, and in parts of East Antarctica, it can represent almost a million years. In order to regenerate those huge stacks, we're looking at tens to hundreds of millennia of ice reconstruction. But despite the sluggish pace of that change, the societal and ecological effects of sea-level retreat will still be profound over the long term. Millennium after millennium, formerly deepwater rocks and reefs will become shipping hazards, maritime towns and port facilities will be repeatedly stranded inland, and near-shore islands will be hitched, one by one, to the mainland by land bridges as new islands arise from the exposed tops of reefs and seamounts.

Although sea-level change will be more of an economic problem for most people than a life-threatening one, it will present much more serious challenges to species that are less mobile or adaptable than we are.

Salt marshes require daily immersions and dryings in order to survive and continue their important work of harboring infant fishes, crabs, shrimp, and shellfish. During previous deglacial periods, salt marshes avoided extinction by moving inland ahead of the tide line, but that's not always an option for them now. We're already holding much of that higher ground. Instead of simply giving them a landward nudge, rising seas are crushing more and more of the world's intertidal habitats against an impenetrable wall of humanity.

Farsighted citizens in south Australia's beachfront city of Adelaide are already struggling to save local salt marshes from a deadly combination of ocean encroachment and land subsidence caused by compaction and groundwater removal. Plans are afoot to remove or reposition barriers such as seawalls and roads that now prevent the marshes from retreating inland, but the specter of long-term sea-level rise casts a deepening shadow over such efforts. Australian coastal scientist Peter Cowell recently expressed his frustration about it to the *Adelaide Sunday Mail,* asking at what point do "we give up and decide we'd be better off relocating rather than trying to defend this stuff?"

Tropical mangrove forests also face similar threats, as do the marine creatures that depend on them and the terrestrial wildlife that has been driven into them by human activity in the surrounding countryside. On the low-lying Bengal coast, for example, dense mangrove tangles make a shrinking haven for endangered tigers, as shrimp farms chew away at their edges even more rapidly than sea level does. Also threatened are intertidal mudflats where commercially and ecologically valuable shellfish grow like root crops, all dependent on narrow ranges of water depth and therefore vulnerable to sea-level change whether it be upward or downward.

Global cooling will flip today's ecological challenges into reverse gear for whatever remains of marshes, mangroves, and mudflat communities in those future centuries. Fortunately, the slow sinking trends will be easier on most coastal ecosystems than the faster warming-driven rises will be.

In the tropics, the crests of shallowwater reefs may gradually die off from increasingly common exposures to the open air as the ocean surface sinks, but new coral colonies might thrive in deeper waters along the reef edges—unless ocean acidification stops them. Reef expert Paul Blanchon, however, finds little evidence of this among the fossil corals of Barbados and New Guinea. "Reefs didn't necessarily die off altogether when sea level fell in the past, but new reef structures didn't grow like they usually do, either," he told me. "I imagine this is because the corals were forced to move downslope and smeared themselves like a veneer over large areas rather than building upward like they did during periods of sea-level rise."

The fastest environmental changes will occur during the next several centuries, while the warming and melting are most intense. And how will people react to those changes? To help answer that question, one might simply look to coastal cities that have already been sinking as a result of human-induced earth movements. You might be surprised to learn how common they are; most of us hear little about their struggles with land subsidence or overlook their similarity to those caused by sea-level rise. But whether you're sinking below the waterline or having the water climb up and over you, the results are much the same.

Venice is one case in point. It is already partly submerged, and it currently drops by twice the rate of sea-level rise as its buildings press into wet muds and supporting groundwater is pumped out. Residents have complained, structures have been lost, and corrective measures have been tried with variable success and at great cost, but it's been more of an aggravation than a deadly disaster. And, of course, the submergence has also helped to turn Venice into a world-famous tourist destination.

New Orleans is also informative. It has long been sinking into the Mississippi Delta as sediments compress and de-water under their own weight and that of overlying buildings, as levees prevent the deposition of regenerating river muds, and as bedrock responds to lingering ice age distortions of the continental crust. In fact, most of the Gulf coast is also subsiding as a result of petroleum and groundwater extraction; some sites have dropped 10 feet (3 m) during the last century. Much of New Orleans itself is sinking by a quarter of an inch (6 mm) per year, with some sections dropping four times faster than that. Clearly, any city that crouches ever farther below sea level faces real danger from powerful storms. But most residents didn't try to move out of New Orleans until hurricanes Katrina and Rita actually rolled in on them in 2005, and many are now rebuilding their homes despite the inevitability of future hurricane strikes.

Much of lowland Tokyo sank 6 to 13 feet (2 to 4 m) during the last century as a result of groundwater extraction; subsidence rates near the harbor can exceed 4 inches (10 cm) per year. Similar processes drive the soft ground beneath Bangkok, Thailand, downward as fast as 4.5 inches (12 cm) per year; that's over twice the speed of Hansen's extreme inundation estimate and forty times the recent rate of sea-level rise. And every year China's largest city, Shanghai, sinks a third of an inch (10 mm) deeper into the Yangtze Delta. During the last century it dropped nearly 9 feet (3 m) and suffered billions of dollars in structural and flood damage.

These examples help to show what the future advance of the sea will really be like in most cases; slow, unrelenting, costly, exasperating, but rarely deadly to humans. It won't be a frothing shoreward rush of waves, but it will still be well worth slowing down as much as possible, as residents of Shanghai and the other already-sinking cities would surely agree. In human terms, global sea-level rise will turn what has until now been a problem faced in isolation by several dozen urban centers into a worldwide phenomenon.

I suspect that our descendants will deal with those coming changes as we have usually dealt with such environmental disturbances thus far—that is, piecemeal. Some who can afford to dike a strip of shoreline or build urban flood-control gates will do so, and some will pull up and move well ahead of the waves. Others will simply accept the risk of being increasingly vulnerable to storm surges, trust in fate, and hang on to the bitter end, hoping that the Big Storm doesn't strike while they're still there. Then as now, mistakes made by those charged with the public welfare will occasionally make climate-driven threats more devastating than they would otherwise be. And when global cooling eventually turns the advance of oceanic waters into a retreat, successive generations will move back seaward and resettle newly revealed lands that were once submerged.

What remains less certain is how these changes will affect coastal habitats and species in the Anthropocene future. They survived even greater shifts in ocean levels during the transitions between ice ages and interglacials by moving inland or seaward as the oceans rose and fell. But now, in a world dominated by humans, we can only hope that there will be room enough on the newly crowded edges of the sea for them to do so again as the shorelines shift beneath them.

8

An Ice-Free Arctic

*The fate of almost everything in the winter
world is ultimately determined by the crystallization
of water.*   Bernd Heinrich, Winter World, 2003

"The polar ice caps are melting!" It is said so often nowadays that it
risks becoming a cliché, albeit one that happens to be true. Arctic ice
retreat is one of the most obvious signs of global warming and it will
drive enormous environmental and societal changes for a long time
to come. But there are also greater depths to the subject than we usually hear about. What, exactly, do we mean by "polar ice caps"? Why
are they melting? Are the polar bears doomed? And what might the
world be like without ice at the North Pole? Would it be bad, good, or
some mix of the two?

For now, we have an ice cap atop each pole, but only the northern
one is likely to vanish within our lifetimes. That's because it is a relatively
thin lid of floating sea ice, fundamentally different from the one in the
south, which rests on solid ground rather than on 2 or 3 vertical miles
(4 to 5 km) of salty water. The seasonally grown stuff that makes up at
least half of it only thickens to 6 feet (2 m) or so over the course of a winter, and the multiyear ice averages about 10 feet (3 m). Even before the
shrinkage of recent decades, the shell of Arctic ice was already so thin in
places that Russian and American submarines used to push right up
through it in order to sneak a look around from time to time.

In contrast, Antarctica is a solid continent trapped under a gigantic
frozen slab up to 3 miles (4.8 km) thick. The world's largest single ice

mass, the East Antarctic ice sheet, is so incredibly cold, thick, and huge that winters on its high austral pole can average 75°F or so below zero (-60°C), far too cold to permit much melting for now. In addition, summer warming (closer to -13°F, or -25°C, at the South Pole) tends to make it snow more in the interior, which thus far cancels much of the loss around the lower, warmer margins.

When we hear warnings about big ice sliding off into the sea, it's the tenfold smaller West Antarctic ice sheet that is most likely to do so. Much of it lies on a peninsula exposed to the additional warming effects of ocean currents, and some parts of it that rest at or below sea level are not firmly anchored. Greenland is also at risk, as we'll discuss in Chapter 9, but it doesn't count as a polar ice cap in the strictest sense of lying on or very close to the pole; the southernmost tip of it doesn't even lie within the bounds of the Arctic Circle.

It's the floating northern cap that we're losing most rapidly, both in terms of area and of thickness. September ice extent shrank by about half between the 1970s and 2006, and in 2007 an even sharper decline reduced it to its smallest extent since satellite-based measurements began in 1979. We don't yet know exactly when this trend will lead to a complete thaw, but most experts expect it to happen well before the end of this century, perhaps even before 2020 AD.

Arctic warming rates far exceed the global average, and most of the circumpolar regions warmed by 4 to 6°F (2 to 3°C) during the last half of the twentieth century. This isn't because greenhouse gases are somehow scrum-piling atop the far north; rather, they're teaming up with other local factors that can also raise temperature themselves, most notably a replacement of reflective white snow and ice by darker, heat-absorbing water, soil, and vegetation. The very presence of so much snow and ice has helped to keep the Arctic unusually cool because much of the energy in the weak northern sunlight is used up in converting water from solid to liquid form, and a great deal of it also reflects off the bright white surfaces without warming them. In a sense, the Arctic is now playing thermal catch-up with the rest of the world because it started several notches lower on the temperature scale to begin with.

But greenhouse warming and reflectivity are only part of the story. Another factor behind the retreat is a natural climate disturbance called the Arctic Oscillation, which makes far northern climates jump suddenly and unpredictably between regionally warmer and cooler conditions. Some scientists blame much of the recent melting on this phenomenon because a switch into mostly positive warm mode occurred in 1989, about the same time that the ice began to retreat. In that mode, stronger westerly winds help to push warmer currents into frozen zones, where they melt the ice from below. The winds also widen the watery gaps between broken floes, thereby allowing more young, thin ice to form and then melt more easily in summer, and they flush multiyear sea ice out into the North Atlantic, leaving more of the thinner seasonal stuff behind.

On the other hand, negative modes of the Arctic Oscillation are supposed to stop those ice-eating patterns, but they've returned briefly several times in recent years without halting the trend, and such reversals have been flopping back and forth for at least a century without devouring the northern sea ice this much. There seems to be something new afoot here and, although it may reflect a joint effort among several causal factors, the artificial greenhouse effect appears to be doing most of the heavy lifting. Even if today's shrinkage slows down again in coming decades, it will probably only delay the inevitable melt-off.

The historical record doesn't provide much comfort in this regard. When the last ice age ended 11,700 years ago, the Northern Hemisphere entered a period in which orbital cycles produced summer temperatures higher than today throughout the Arctic. During that early Holocene warm phase, which lasted roughly 2,000 to 3,000 years at most sites, summer air temperatures were generally 4 to 6°F (2 to 3°C) warmer than now and northern sea surface temperatures were two or three times higher than that. When I asked Johannes Oerlemans, professor of meteorology at Utrecht University, if that was enough to uncover much of the polar ocean, he replied, "It's hard to imagine that summer temperatures several degrees higher than today would not have had a significant effect on the sea ice cover. Many of my colleagues have expressed that opinion, so it's really nothing new."

According to a recent study headed by Darrell Kaufman of North Arizona University, the remains of blue mussels, *Macoma* clams, and bowhead whales in marine deposits show that these animals invaded much of the Canadian Arctic Archipelago and Beaufort Sea coast during the warm early Holocene in response to newly opened waters. Similar evidence from farther east also reveals similar conditions in and around Svalbard, but those conditions were not perfectly representative of what is happening now. For one thing, that earlier insolation-driven warming was primarily limited to summers at high northern latitudes rather than the more universal effects of greenhouse gas buildups that we face today. In addition, winds and ocean currents and temperatures in the eastern Canadian Arctic were still strongly influenced by what was left of the gigantic Laurentide ice sheet. In light of those differences, we need not be surprised to see even more dramatic sea ice retreats despite somewhat lower temperatures today.

And there's yet another point that is often overlooked when the subject of polar melting arises in the media. When most experts say that the North Pole will soon be ice-free, they're not talking about total, year-round loss. Both poles endure long months of darkness every winter and temperatures fall far below zero then. Even now, the Arctic ice cap doubles in size as the December–January air chills down, and all but the most extreme versions of future warming might not keep all of it from refreezing during the dark months of winter. This is mainly about melt-backs in late spring and in summer, when the circling Arctic sun warms the pole twenty-four hours a day.

Most of what we hear about the northern ice retreat is that it threatens polar bears. Sometimes we also hear that a new Northwest Passage has opened along the Arctic coast of Canada and that shipping, mining, and fishing interests are moving in to exploit the new geography. Such broad statements leave a lot unsaid, but together they raise an important point: some will lose and some will gain as the Arctic opens up. And time will also change that list of losers and winners when the dramatic transitional events that hold our attention now are over and an ice-free polar ocean begins to seem normal, then slowly begins to glaze over again. By digging deeper than usual into such details of life in the far north, I hope to provide some helpful insights into what is going on up there as well as some of what's to come.

Polar bears are an obvious subject to start with, but even the simplest online search yields a confusing mess of opinions about what is really happening to these iconic symbols of global warming. Some articles tell of bears drowning in a desperate search for ice to stand on, echoing the computer-generated scene in Al Gore's film, *An Inconvenient Truth*. Others excoriate such reports for being misleading; the Danish statistician and author Bjørn Lomborg begins his book *Cool It* with one of the stronger critiques, calling most versions of the story "vastly exaggerated and emotional claims that are simply not supported by data."

Most of these reports apparently stem from an incident in 2004, when scientists who were flying off the coast of Alaska spotted several dead bears floating many miles from the nearest floes. A powerful storm had just struck the area, and the animals were probably overwhelmed by violent waves while swimming in open water. The ice margin lay farther offshore than usual that year, and bears had already been seen swimming more than 50 miles (80 km) from land. Normally, a long-distance swim isn't particularly troublesome for a polar bear because the adults are comfortable enough in deep, cold water to qualify as honorary amphibians. However, the recent ice retreat makes such journeys more common, so the risk of being caught out in a dangerous storm is probably increasing, too. Furthermore, the smaller body sizes of young cubs make them more vulnerable to hypothermia in cold water, so longer swims may be a more serious problem for them than for their parents. For now, the significance of this issue remains inconclusive; yes, bears have drowned, but how many are actually dying because of climate change?

Keeping in mind the highly politicized nature of the subject, I have used extra caution in retrieving what I believe to be reliable background facts and updates from the field. My principal source is Andrew Derocher, a renowned polar bear specialist who is based at the University of Alberta and who kindly agreed to talk with me by phone.

That was no small favor. As one of the few real experts in a field that draws a lot of attention these days, Derocher is besieged by journalists and climate naysayers in search of a catchy quote or a punching bag. When I called him, he had just finished reporting a particularly threatening letter to the campus security office.

For many of us, the north polar cap can seem about as appealing as an empty parking lot in winter; nothing but lifeless, featureless white emptiness. But Derocher sees it differently. "Under normal conditions," he explained, "you get huge pressure ridges of ice that stretch for miles and act like snow fences that trap powdery drifts on their leeward slopes. Ringed seals come up through cracks in those ridges and dig birthing lairs in the soft snow pockets."

It's the seals that draw bears out onto the ice in the first place. Unlike their southern cousins, polar bears generally stay awake and hungry even in the long darkness of winter, although pregnant females do spend that season in snow-covered maternity dens where their cubs are born. Meanwhile, nonmaternal adults prowl the drifts and ice ridges in hopes of snagging the odd unlucky seal, but they usually have little success until the light returns and their favored prey's whelping season begins.

Seals sustain the bears, and ice sustains the seals. As the sea ice shrinks in the warming Arctic, the effects cascade through the food web. "Ringed seals establish their winter territories in late autumn, when the ice sets up for the winter," Derocher continued. "As the freeze-up comes later and later, seals are less and less likely to find their territories in time, and that makes them less likely to den up and give birth in spring." Ringed seal behavior, it seems, is closely tied to day-length cues, which remain pinned in place on the calendar, while the duration of winter ice

Polar bear on Hudson Bay sea ice. *Andrew Derocher*

cover shrinks at both ends of the season. When breeding time arrives on schedule, the seals still heed the call, but Derocher fears that they may be forced to stake their territorial claims on more dynamic, shifting offshore ice that makes the birthing and raising of pups more difficult.

When waxing spring sunlight heralds the arrival of pups, the bears have two or three months in which to do most of their feeding for the year. They sniff out buried dens and pound, stiff-legged, on the snowy roofs with their forepaws, hoping to grab the youngsters before they scoot down their escape chutes into the water. Perhaps one in twenty attacks suceeds, but that's enough; ringed seal pups are incredibly energy-rich snacks, with fatty blubber making up half of their body weight. But it's not just the pups that they're after. "Females with pups make bigger meal packages," Derocher explained. "A bear will sometimes discard a pup and then wait at the birth lair for the mother to come back."

Weeks later, the southern lip of the ice peels slowly back toward the pole and most of the bears move ashore. In western Hudson Bay, where Derocher does much of his research, summer is a lean time even though some terrestrial foods are theoretically available. From May through August, the Hudson Bay bears usually fast and wait for the coastal waters to refreeze. Pregnant females there may go without eating for as long as eight months before seeking shelter for the winter. In spring they emerge from their maternity dens and move back out onto the returning rim of ice with their new cubs.

But not all individuals are alike, and different habits develop among different subgroups. Polar bears in some regions spend more time on land than polar bears in other regions. Some may take a break from sealing to hunt walrus and narwhals offshore and then stalk reindeer and geese onshore. In the Svalbard archipelago, north of the Norwegian mainland, bears have been seen chewing on beached sperm whale carcasses. In western Hudson Bay, some catch ptarmigans or munch berries. And, on rare occasions, human flesh goes down easily when the belly is growling.

It can be tempting to hope that innate resourcefulness will help at least some polar bears to survive future warming. But Derocher cautioned, "they certainly are adaptable, but there are limits." The main problem is their need for high-calorie meals. Foods other than blubber are too energy-poor to keep a bear in good physical condition through four- to eight-month fasts. It was probably this narrow ecological niche, the hunting of seals on pack ice, that originally led these predators to evolve into white, water-loving versions of terrestrial brown bears.

Polar bears split from their brown/grizzly relatives roughly 200,000 years ago, perhaps when an ice age marooned a group of ancestors in some glacier-rimmed pocket of the Arctic. The length of that lineage overlaps the dates of the Eemian interglacial, so they've endured at least one long-term warming before. When an Icelandic scientist recently discovered an Eemian-age polar bear jaw bone in Svalbard, an article posted online by the World Climate Report quoted him as saying, "the polar bear has already survived one interglacial period. So maybe we don't have to be quite so worried about them."

Derocher dismisses such comments as misleading. "To me, all it really means is that enough sea ice persisted through the Eemian to support polar bears." Although some evidence suggests that the last interglacial saw open-water summers at the North Pole, we still have too few sediment core records from the Arctic Ocean to confirm or refute the idea of a total melt-off, and Derocher's hypothesis may well be correct. He continued: "During the last ice age, they lived as far south as Germany and southern Scandinavia, but when it warmed up again they abandoned those areas even though some ringed seals still persist in the Baltic today. The bears probably can't hang on for long without the stable, extensive sea ice that their hunting behavior centers on."

If Anthropocene warming eliminates spring and summer sea ice altogether, there will be no northern refuges left to escape into. These supreme maritime predators will have to adapt to hunting on land or perish. That, of course, could be difficult if your coat is white on hunting grounds that are the color of dirt and leaves, and the only edibles available, from the point of view of your seal-specific digestive system, are the nutritional equivalent of junk food.

Meanwhile, brown bears will be moving northward in response to warming and the advance of boreal forests over tundra. What will happen when the two closely related species meet again? One "polar-grizz" hybrid with a brown-spotted white coat and a grizzly-like shoulder hump has already been documented (that is, shot) in Canada's Northwest Territory. Perhaps such half-breeds might become more common in the future. More likely, though, the polar refugees will be outcompeted by their omnivorous relatives who already feel more at home in terrestrial habitats.

To Derocher, the writing is on the wall. As the ice in western Hudson Bay melts back earlier in spring, it gives the bears less time at the annual seal feast. And the ringed seals are feeling the effects of warming, too. "As the sea ice thins and weakens, you don't see as many pressure ridges forming. Instead, it just rafts up into rubble piles that don't capture drifts, and the seals are beginning to use ice chambers that aren't as well insulated as fluffy snow lairs are. Their escape holes can sometimes freeze over now, so the mothers have to break through them to get to their pups, and sometimes they might not even find the entrances." This change in den architecture also affects the bears. Derocher has recently seen some of them try to scrabble and scratch through solid ice in pursuit of denning seals, usually with little success. "It's a new hunting method that's a lot different from punching down through soft snow, but it's not so much a matter of innovation on their part," he said. "It's more like desperation."

The less a polar bear eats, the less it weighs and the less energy it can dedicate to reproduction. "It all boils down to body condition," says Derocher. "Females who weigh less than 190 kilos are far less likely to breed successfully. Mothers who weigh more than that but are still underweight produce fewer cubs, and those cubs are born smaller." In a world where body fat not only sustains you but also insulates against frigid winds and waters, the trend toward skinnier moms and fewer, smaller cubs points in only one depressing direction.

The main problem with the white bears, then, is not that they're all drowning, or even starving to death. They're just not breeding enough. It is the species itself, not so much the individual animals, that is most in danger now, especially in the southern limits of the Arctic where the effects of warming are felt most strongly.

Finding the numbers to support that claim, however, is no easy task. Polar bears cluster into more than a dozen circumpolar populations, and the few scientists who census them under difficult field conditions put their total numbers between 20,000 and 25,000, with about two-thirds of the animals living in Canada and perhaps 3,000 hanging out in the Barents Sea region. Meanwhile, a blizzard of controversy rages over whether those numbers are rising or falling as the ice cap shrinks.

Reports of increasing numbers are probably accurate in some locales, but some may also stem from recent improvements in census methods that allow more individuals to be found, and also from certain parties with vested interests in promoting bear hunting or denying climate change. In fact, the regionally comprehensive data needed to confirm large-scale trends of any kind are still lacking. In western Hudson Bay, though, the data are in; the population declined from about 1,200 animals in 1987 to 935 in 2004. According to Derocher, nutritional effects on reproduction are the most likely culprits, and sea ice retreat is the most likely cause.

No such slimming has yet been found among the Svalbard bears on the other side of the pole, but that's not surprising. The ice in the Barents Sea is more extensive than it is at warmer, lower latitudes, and those bears still spend most of their time offshore bothering the seals. Most of the other subpopulations are too sparsely studied for us to be sure how they're doing.

And what of the future? Some polar bears are beginning to eat more bearded seals and harbor seals as the ringed types become harder to reach from shore; those alternative prey will probably become more abundant while the favored sea ice habitats of the ringed seals retreat poleward, and an increase in harbor seal populations is already occurring in Hudson Bay. Perhaps the predators will manage to hang on this way in the more remote corners of the north.

But Derocher worries when he considers the future, including his own. "I used to think that I might not live to see these changes," he said. "Now I'm close to fifty years old and it looks like they'll happen before my career is over. I try to be optimistic about it, but the thought of spending my last productive years helicoptering around to document the demise of the polar bears isn't very appealing."

He will avoid that fate only if a lot of "ifs" fall into place. If ringed seals alter their breeding behavior appropriately, and if changing marine communities provide them with enough food, and if bearded and harbor seals can be as nutritious and attractive to bears as ringed seals are, and if landlocked polar bears can avoid competition and interbreeding with brown bears by taking refuge in northern Greenland or Svalbard, and if expanding northern industries and human populations don't harm the wildlife much . . . then, maybe, we'll still have polar bears with us when the Arctic Ocean finally refreezes at some later date in the deep future. The recovery from a moderate emissions scenario drops us back near today's temperatures in a few tens of thousands of years, and an extreme scenario does it in several hundred thousand. But that's a long time and an awful lot of "ifs," and it's hard to justify putting polar bears and ringed seals anywhere but on the list of probable losers as the Anthropocene runs its course.

Sadly, they're not alone on that list. Walruses are also in jeopardy as their nursery habitats melt away. Walrus moms leave their helpless young behind to wait on sea ice platforms as the adults grub around for bottom-dwelling clams and crabs with their bluff, bristly snouts. But as the ice margins retreat farther and farther offshore, the water beneath them grows ever deeper and diving takes more time and energy. Walruses are large and powerful, but they don't have gills and they can't swim indefinitely, so they have to pull out and rest between dives. When water depths exceed 650 feet (200 m) or so, it's just too energy-consuming to continue, and the adults have to go elsewhere or starve. Apparently, they sometimes abandon their infants in the process.

Nobody knows how widespread this problem is yet, but one heartbreaking set of field observations got a lot of media coverage recently. In 2004, a U.S. Coast Guard icebreaker encountered nine baby walruses swimming alone in deep, ice-free Canadian waters. Shipboard biologist Carin Ashjian told *Science Daily*: "We were on a station for twenty-four hours, and the calves would be swimming around us crying. We couldn't rescue them." Like other ocean-dependent mammals of the north, walruses will have to learn new ways of balancing the demands of foraging and child care if they're to survive the loss of summer ice. "The young can't forage for themselves," Ashjian continued in her interview. "They don't know how to eat." Instead, they live on mother's milk for up to two years. But not if Mother is nowhere to be found.

For many of us who only experience the Arctic indirectly through the media, such a focus on mammals dominates our ideas about what lives up there. But there's a lot more going on beneath the ice than above it, and it's as much at risk as the large charismatic creatures topside.

Brine pockets in the sea ice form a porous network of channels that algae and other microscopic life-forms thrive in. The translucent ice also transmits enough spring and summer sunlight to sprout rippling meadows of algal filaments on the frosty undersides of the floating roof. Native species of tiny, shrimplike copepods wander those inverted green, white, and blue vistas to graze on the hanging gardens. Further up the food chain, Arctic cod dart in and out of sheltering crannies in the ceiling. Native Arctic whales, the ivory-white belugas and unicorn-tusked narwhals, are also closely linked to floating ice. They lack the prominent dorsal fins of many open-water whales, and that makes it easier for them to heave broken floes aside when they need to breathe and chase fish around the underwater icescapes.

The wide, shallow continental shelves of the Arctic Ocean are also very much alive, and some of the world's densest aggregations of bottom-dwelling creatures thrive on organic detritus that rains down from the upper waters. Slithery brittle stars can number in the hundreds per square yard, and prickly sea urchins, cold-tolerant clams, fat sea cucumbers, and wriggling polychaete worms support a wealth of predators, from fish to birds to mud-grubbing walruses.

These highly specialized communities are now disappearing along with the ice. Species from lower latitudes are moving into the warming, opening waters, a process that some biologists call the "Atlantification" of the Arctic, with harbor seals, harp seals, and Atlantic cod among the most numerous of the immigrants. Meanwhile, whale-hunting orcas are also invading the newly ice-free waters. Their high dorsal fins make it difficult to navigate overhead ice cover, but an increasingly navigable polar ocean is leaving belugas and narwhals more vulnerable to orca attack. The calves of bowhead whales also make tempting targets, and the adult bowheads also face increasing competition from minke whales that are moving in on their feeding grounds.

Less visible threats face the native northern whales, as well. As southern animals follow the retreating ice northward, their microbes come with them, and pilot whales and their smaller relatives carry distemper, brucellosis, and other diseases that circumpolar belugas and narwhals currently lack resistance to. Although diseases are unlikely to exterminate entire species, adding epidemics to the list of new environmental pressures isn't a pleasant prospect. Canadian marine mammal expert Otto Grahl-Nielsen recently estimated that half of the Arctic's belugas and narwhals could eventually perish from imported distemper alone.

To be among the winners in the new Anthropocene Arctic, it now helps to be a generalist open-water species rather than an under-ice specialist. Even the tiny copepods are responding. Indigenous species with place-based names like *hyperboreus* and *glacialis* are giving way to immigrants that don't need the hanging gardens of ice algae. Open-water fish such as pollock and salmon are replacing species adapted to frozen habitats, and aggressive Atlantic cod are displacing or eating their smaller polar cousins. In the Canadian Arctic, thick-billed murres, which resemble miniature, fast-flying, northern versions of penguins, are beginning to feed their chicks with capelin, a small plankton-eating fish that prefers waters with little or no summer ice, rather than the slightly larger Arctic cod that dominated their diets until the mid-1990s. Scientists who study seabirds also expect North Atlantic puffins and common murres to follow their traditional capelin prey northward as well, and razorbills (which look a bit like murres) are beginning to colonize rocky islands in the Hudson Bay region.

Judging from the changes already under way, life in an ice-free Arctic Ocean will probably be a mixture of whatever survives the de-icing and what moves in from the Atlantic and Pacific. But unsheathing the waves under the Anthropocene sun will also be like switching the lights on in a fertile greenhouse. Neil Opdyke, a noted authority on oceans and climates at the University of Florida, is one of many experts who foresee a burst of planktonic marine life ahead. "It's going to be a very different and very productive ecosystem," he speculated with me recently over lunch in Gainesville. "Not only will you have the ice-free ocean, but the thawing of permafrost will flood rivers with soil nutrients that will eventually end up offshore."

That combination of warmth, light, and nutrition will trigger population booms of floating microalgae, or phytoplankton, and longer growing seasons and reduced ice cover have already boosted Arctic phytoplankton abundances during the past decade. Many of the world's most productive fisheries develop on just this sort of ecological sweet spot, where plentiful light and nutrients support immense amounts of microbial growth. Such sites include upwelling zones along the coasts of Peru and Namibia, as well as the storm-stirred Southern Ocean. Phytoplankton, in turn, feed shrimplike krill and copepods, which support vast schools of plankton-eating fish. Then come the predatory fish, the gulls and puffins, and the orcas and harbor seals.

While Arctic cod and belugas fade away, the ice-free pole may open a new, food-rich refuge for other species that now face overharvesting and pollution problems farther south. With so much biomass production going on, enterprising humans will surely try to cash in on the new source of marine protein as well, with an open sea to harvest it from and an increasingly habitable coastline to support the industry. An online article posted by the Euroarctic.com news service in 2006 reported that Russian companies are already building specialized trawlers to harvest anticipated bounties of fish from the polar ocean.

But we're forgetting something that could trouble these waters even if the whole place could be designated a hands-off sanctuary. Carbon pollution not only warms the oceans, it acidifies them, and Arctic marine organisms will be among the first to feel the corrosive effects of those acids. Even in a moderate 1,000-Gton emissions scenario, the chalky aragonite shells of many organisms, from coccolithophores to clams, will soon begin to dissolve in acidified polar waters. The coming one-two punch of climatic and chemical shifts will produce biological "no-analog" situations unlike anything seen during the Eemian or early Holocene, the nature of which even the most capable marine biologists can only guess at.

What might life be like in an ice-free and *acidified* Arctic Ocean? Total algal productivity probably won't suffer much; most marine phytoplankton don't build soluble carbonate coverings. The favored microalgal forms will probably be golden brown diatoms, whose sparkling transparent shells are made of acid-resistant silica rather than carbonates. Diatoms already thrive in the northern Atlantic and Pacific, and they'll probably do well in the ice-free Arctic, too.

Among the invertebrates, potential losers might include shell-bearing pteropods, forams, mollusks, barnacles, urchins, crabs, and cold-water corals. Winners could include soft-bodied sea angel pteropods, jellyfish, anemones, sea cucumbers, and worms. Whether an ecosystem dominated by soft-bodied winners rather than crunchy losers is good or bad is a matter of personal taste, and it's therefore judged differently in different situations. We might grimace at the thought of swarming schools of jellyfish at the North Pole, but such swarms are an awe-inspiring tourist attraction in the Pacific island nation of Palau.

In fact, nobody knows for certain which species will persist or perish in that unfamiliar new world. The rich genetic complexity of life provides abundant material for natural selection to act upon and many marine organisms survived the acid bath caused by greenhouse gas buildups during the PETM superhothouse of the early Cenozoic. But we do know that major ecological changes are coming to the Arctic Ocean, not just in terms of climate but also of pollution, commercial development, and exploitation of natural resources, and that they will involve much more than polar bears alone.

The changes on land will be dramatic, too. Boreal forests are already spreading northward over former tundra, and tundra is creeping over formerly barren polar deserts. By some estimates, a large expansion of northern vegetation could partially offset some of the future greenhouse gas buildups by sequestering carbon in plant tissues. In certain regions, woodlands may meet the encroaching sea; in others, sodden bogs will hold the trees at bay. Native birch forests are expected to replace mossy tundra in much of Iceland, and pines are set to invade northern Sweden and Norway. Tundra flora in general are more likely to be harmed than helped by warming because they can't move any farther north, though some may find refuge in colder Arctic highlands.

Along with the new plants, new animals such as moose, mink, red foxes, wood-boring beetles, and butterflies are also moving to higher latitudes. In Canadian rivers and lakes, southern brook trout and introduced brown and rainbow trout are likely to move north as well, perhaps displacing native char. Ironically, this means that the total biodiversity of the Arctic is increasing. It shouldn't be surprising, considering the well-known pattern of higher species counts at warmer latitudes, and it might be thought of as beneficial were it not linked to the decline of unique species and were it not caused by human-generated pollution. However, little or nothing in terms of net global-scale biodiversity is gained by the change because the lengthening list of Arctic residents contains no organisms that are new to the planet as a whole. Most of the immigrant "winners" that are moving into the increasingly enriched communities of the warming Arctic are already well established farther south, but many of the "losers" have nowhere else to go and may be lost to extinction.

Some of the likely losers among terrestrial Arctic mammals are caribou, reindeer, pika, and lemmings. Competition with newcomers for food and transmission of parasites and diseases from invasive species are only some of the problems that they face; warming itself is yet another. More frequent cold-season rains and thaws destroy the protective insulation and oxygen flow necessary for winter-active rodents to survive in their under-snow tunnels. They also encase edible mosses and lichen with impenetrable ice, and hard crusts within and upon formerly loose snowpacks make travel and digging for food more difficult. A single October rain-on-snow event killed 20,000 musk oxen on Banks Island in the Canadian Arctic in 2003. On the other hand, longer and warmer growing seasons might make life easier on those animals who make it through the cold months.

Warming even reshapes the land itself. Where smooth bedrock underlies seasonally frozen water bodies, the lengthening of summer open-water periods is driving some of those lakes into oblivion. My friend John Smol, a specialist in the history and ecology of lakes at Queen's University, Ontario, is now losing some of his study sites to a process that he calls "crossing the final ecological threshold" of human impact.

For thousands of years, shallow depressions in the glacially smoothed granite of Cape Herschel on Ellesmere Island have contained frozen ponds that thawed only briefly during summer. In an earlier study, Smol and his colleagues used sediment cores to show that the ponds were becoming ice-free for longer and longer periods since the nineteenth century; sunlight-loving algae were common in the most recent layers of mud but rare or absent in the older ones. They were criticized at first for being alarmist, but the trend has now become undeniably clear. "The longer these lakes stay ice-free in summer," he explained to me, "the more water they lose to evaporation in the twenty-four-hour sunlight. Some of them have now dried out completely, and the rest are heading in that direction, too."

Scientists from Queen's University, Ontario, collecting the last samples from what used to be a pond at Cape Herschel, Arctic Canada, now drying out as a result of recent climate change. *John Smol*

I asked John if he has noticed other signs of change, as well. "Sure," he replied. "Native ice-dependent species are losing out, but lots of other stuff is moving up from the south. We're now seeing robins on Baffin Island, and bees are showing up on the tundra flowers. The Inuit have no traditional names for some of these new species, and they're even having trouble telling when to harvest the local berries because the growing seasons are changing so much." Among the more unpleasant differences is the arrival of biting insects. "Our study site is so far north that we never had much trouble with mosquitoes before; it was just too cold up there. But now we're starting to run into them in the field." One win for the bugs, one more loss for the scientists.

Is he depressed about these changes, or just "scientifically concerned" about them? "Depressed," he answered without hesitation. "But not quite despairing, either. I still rage about what's going on, but that's because I still have hope that we can stop things from getting a lot worse."

Elsewhere, hard permafrost soils are thawing, turning vast stretches of tundra into quagmires. Such places are seeing new meltwater ponds appear, quite the opposite of the situation at rocky Cape Herschel. And much of the northern coast is low-lying, so newly ice-free ocean waves cut deeply into soft, formerly protected shores. Few maps of the expected encroachments of sea level on the world's ports and beaches show the circumpolar regions clearly, so it can be difficult to envision exactly what this means for Arctic geography. But it appears likely that the changes will be large, especially where loosely consolidated land slopes most gently upward from the sea.

Horrifying examples of this process are already under way, not so much from rising sea levels but simply from the warming of frozen soils and the loss of protective coastal ice. Many native settlements along the shores of northern Alaska and Canada were built upon permafrost-cemented sediments close to the beach, but what was once a firm foundation is now becoming loose and waterlogged. That is a serious problem in and of itself; some folks now have to wait until nocturnal cold solidifies local roads before traveling between villages. But in the face of an unsheathed ocean, the softening earth also becomes easy prey for the hungry waves, particularly when storm winds pile them high and heavy against the shore. Parts of the Beaufort Sea coast retreated by 43 to 46 feet (13 to 14 m) per year between 2002 and 2007, dumping exposed villages and archaeological sites into the water from steep, crumbling permafrost bluffs.

As disturbing as they are to watch, these scenes of buckling roads, house-eating potholes, and collapsing coasts are temporary signs of a polar world that is currently in transition. The rising surface of an open Arctic Ocean—and, much later on the Anthropocene carbon curve, falling sea levels—may keep vulnerable coastal settlements in a chronic state of flux for centuries. But eventually much of the water that now soaks the spongy surfaces of peatlands may evaporate or drain away, allowing the land to settle into a drier, more stable condition that can better support forests, fields, roads, and towns. Shipyards, refineries, and storage facilities will turn formerly remote outposts into busy transport hubs and ports of call. Pole-crossing marine shortcuts are already being mapped between Iceland and Alaska and between Murmansk and the Hudson Bay settlement of Churchill, which is now linked to the rest of North America primarily by air and rail.

As polar habitats continue to evolve, more people will move up there as well, especially where new industries take root. As traditional indigenous hunters of caribou and whales face the loss of familiar patterns of weather and ice stability, others will aim to join the list of winners, and not all of them will be outsiders. *Nature* correspondent Quirin Schiermeier recently quoted Frank Pokiak, chair of the Inuvialuit Game Council in Tuktoyaktuk, Alaska, as saying, "People need to understand that we've been living with changes all our lives. Climate change is just another thing we have to adapt to. We may need to harvest other species, perhaps grizzly bear, perhaps caribou, but we won't quit existing." Aiding such people in their adaptive efforts will be a host of warming-related factors, from lowered heating costs to easier open-water travel to longer and more job-rich construction seasons.

With clear passages more and more reliably open in summer, more ships will sail the coastlines of the polar ocean. Unfortunately, not everybody agrees upon who owns what in that new frontier, and territorial disputes are heating up even faster than the air overhead.

A case in point is Hans Island, a wedge-shaped half square mile of stone in the narrow strait between Greenland and Ellesmere Island. Both Denmark and Canada consider these to be private territorial waters, and their originally claimed boundary lines overlap to an uncomfortable degree. In the past, such differences of opinion didn't threaten much more than nationalistic pride, but the stakes are higher now.

A tit-for-tat struggle over the ridiculously tiny islet commenced during the 1980s when a Canadian oil company began to reconnoiter the area. In response to the perceived challenge, the Danish minister for Greenland helicoptered over to Hans Island to plant a flag along with a bottle of liquor and a note that reportedly read "Welcome to the Danish island." In the years that followed, alternating military patrols of Canadians and Danes took turns pulling each other's flags down and erecting nationalistic markers of their own. Though amusing, this dispute was also potentially quite serious. In 2005, Canada's Conservative Party called the latest Hans Island incident "an invasion of Canada," and another politician threatened to use naval vessels to "protect our territorial integrity" up north. The flag war finally ended peacefully when precise satellite imagery convinced officials in Canada and Denmark that the international border actually splits the island rather than sidestepping it.

The increasingly ice-free Arctic is a potentially valuable new resource for northern nations. The expansion of shipping lanes into the polar ocean, with open water in summer and only thin seasonal ice in winter, will revolutionize the geography of Russia, Canada, Scandinavia, and Alaska. The newly opened Northwest Passage and today's patchwork of isolated "seas"—the Laptev, Beaufort, and Barents Seas, the Chukchi and East Siberian Seas—will fuse into one continuous water mass, and the pole, once a barrier, will become a gateway. Ships hauling cargo and passengers between Europe and Pacific nations will bypass the Panama Canal for much or all of the year, depending on how iceworthy the ships are. A Northwest Passage route from Rotterdam to Seattle is 2,000 nautical miles shorter than the current one through Panama, and tracing a "Northern Sea Route" along Russia's coast rather than using the Suez Canal cuts 4,700 nautical miles from the trip between Rotterdam and Yokohama.

Some in Canada are trying to claim the Northwest Passage for their homeland by citing the 1982 UN Convention on the Law of the Sea, which pushed national boundaries 200 miles (320 km) offshore and therefore may put most of that lucrative coastal route within the economic sphere of Canada. Russia and Norway are arguing over rights to the Barents Sea, and Russia, Canada, Norway, and Denmark are all trying to stretch their northern claims as far offshore as possible. Most nations seem willing to leave the central high seas free and open to international navigation, but it's what lies beneath the waves that's becoming most problematic.

In June 2007, Russian geologists returned from studying the Lomonosov Ridge, a narrow fence of faulted seabed that splits the deep central basin of the Arctic Ocean between Siberia and Greenland. Normally, such an expedition would go unnoticed by the general public, but this one ignited a political firestorm.

The flash point was the geologists' claim that the ridge is, geologically speaking, a direct extension of the Russian mainland. In an instant, long-standing perceptions of the North Pole as being a landless realm of floating ice evaporated. There's plenty of land up there after all; it's just that it lies under several vertical miles of frigid seawater.

Two months later, a Russian minisub took territorial disputes to new heights—or depths—by planting a rustproof titanium flag on the seafloor beneath the pole. Was it merely a prank, or did it amount to claiming the top of the world along with roughly half of the Arctic Ocean bottom? Canada and Denmark were neither amused nor convinced; the other end of the Lomonosov Ridge joins their own territories of Ellesmere and Greenland. At the time of this writing, the Russian claim still awaits confirmation or rejection by the United Nations, which faces a lengthening list of similar territorial dustups.

That so many parties are caught up in a polar land grab attests to the reality and importance of climate-driven changes in the Arctic. Some may still deny that global warming is upon us, but they're certainly not among those who are now maneuvering to cash in on it. The whole situation seems rather unfair, though. Most of the nations that are slated to reap the greatest rewards from this new north are already among the world's wealthiest, and they're also among the carbon emitters who are most responsible for these changes to begin with.

The bedrock north of Great Slave Lake in Canada's Northwest Territories is thought to represent one of the richest mineral deposits on Earth. The Ekati and Diavik mines already produce more than 10 percent of the world's natural gem-quality diamonds and even as northern access routes that depend on stable permafrost and lake ice crossings liquify, engineers and industrialists are envisioning a future spiderweb of new dryland roads and tracks to connect new mines to new marine ports. Both Canada and Russia are preparing for further intensification of a diamond-driven "cold rush" once the roads can support it more effectively. Baffin Iron Mines Corporation is reportedly developing its own Arctic railway and port to service its operations, and uranium miners in Nunavut are poised to spark a new northern "glow rush."

Published estimates vary, but between a tenth and a third of the world's untapped oil reserves are thought to lie in the Arctic, particularly on the broad, shallow continental shelves. There may also be far more natural gas there than oil. North America's largest oil fields lie in Alaska's

Prudhoe Bay, huge quantities of gas and coal underlie the North Slope, and Canada has major reserves in the Mackenzie Delta and elsewhere in the high Arctic. Russia claims valuable deposits along its Siberian shores; close to three-quarters of its oil and gas already come from its Arctic territories. Reliable open-water routes will make those resources more accessible and profitable. They will also increase the risk of oil spills, which could be horrendously difficult to clean up if the black goo slips beyond reach under what remains of the floating ice.

The new fossil fuel bonanza could also drive the sum of our greenhouse gas emissions even higher than it already is, but convincing people to shut those wells down in order to stop the climatic changes that bring them so much wealth will be difficult. Even in countries that currently express interest in controlling carbon pollution, the temptation of short-term monetary gains from Arctic fuel production and trade may outweigh concerns about climate among those who stand to benefit from them.

Ironically, we who brought this new world into being will also be responsible for its eventual demise. As atmospheric CO_2 concentrations finally sink back closer to today's levels thousands of years from now, the Arctic Ocean will begin to freeze over again. But when? Sediment cores collected from the Lomonosov Ridge by an international team under the leadership of Norwegian researcher Catherine Stickley reveal the first appearance of ice-loving algae in layers that date back to 47 million years ago. Atmospheric CO_2 levels are thought to have fallen to 1,100 to 1,400 ppm by then from their earlier Eocene highs, so perhaps 1,100 to 1,400 ppm represents an approximate threshold to Arctic ice formation in the darkness of winter. In that case, the slow recovery from an extreme 5,000-Gton carbon emissions scenario might permit winter sea ice to re-form between 2,000 and 5,000 years from now, and a moderate scenario might not stop winter ice formation at all.

But allowing the Arctic ice cap to persist in summer as well would be another matter. Several sources report that permanent year-round coverage of the pole began 14 to 10 million years ago, when CO_2 concentrations were similar to those of today. If we take the modern value of 387 ppm CO_2 as an approximate benchmark for the transition between frozen and ice-free conditions in summer, then a relatively moderate emissions scenario could delay year-round sea ice cover to some time between 50,000 and 100,000 AD. And by this same reasoning, an extreme 5,000-Gton scenario might delay it for as long as half a million years.

From our perspective in a time period when solid ice cover is still considered to be a normal feature of the Arctic Ocean, it is easy to think of the temporary opening of the ice cap as a shocking environmental wound inflicted by our carbon pollution. But for countless generations of people who will grow up accustomed to ice-free conditions in the deep future, the situation will probably seem quite different. Far out on the fading tail of the Anthropocene's carbon pollution curve, they may well shudder at the incremental shrinkage of the polar sea lanes and the gradual demise, species by species, of what will by then have become familiar, even ancient marine ecosystems. Those ecosystems will have become unique in their own right, having developed in the context of the future Arctic's unique combination of open waters and long, alternating stretches of seasonal darkness and light. Traditional native skills of living on an icebound sea may also be long lost to history by then, so life in a fully frozen north might seem undesirable, if not impossible, even to the descendants of today's seal and whale hunters.

And later still, perhaps some time close to 100,000 AD if we follow the moderate 1,000-Gton emissions scenario, or several times later than that in the more extreme case, the polar seas will lie, like Snow White, encased once more in a glassy tomb. The once-thriving Arctic fishery will fade, along with the Northwest Passage, back into the realm of dreams. And great white bears, if any remain then, may once more wander the spring snowdrifts, pressure ridges, and floes in search of seal blubber. One can only hope that the ringed seals will also be there to share the floating ice with them if they ever do return.

9

The Greening of Greenland

Han kallade landit, þat er hafde fundit, Grænland,
þuiat han kuat, þat mundu fysa
*   menn þangat, er landit heti vel.*

—*Eirik's* Saga Rauda *(transcription by*
Aaron Myer, Northvegr Foundation)

He named the land that he had found "Greenland"
because, he said, it might draw more people there
*   if it is well-named.*   —*translation by Curt Stager*

In 986 AD, Eirik the Red sailed from Iceland as the leader of more than two dozen boatloads of Norse colonists. Their destination: Grøn-land, an indented rocky coast whose thin soils supported enough grass and shrubby browse to keep small herds of sheep, goats, cattle, pigs—and their owners—alive. Eirik had recently killed several of his countrymen over property disputes, and for him the journey was more than a quest for adventure; it was both a banishment and a chance to redeem himself with a successful economic venture. He'd had little trouble convincing several hundred farmers and their families to join him in the settlement effort, thanks in part to a recent famine, the relatively poor quality of most Icelandic soils, and the appealingly lush-sounding name that Eirik himself had given to the new country. The open-water trip was also more attractive than it might otherwise have been because a centuries-long spell of natural warming, what many historians have called the Medieval Warm Period, kept the route unusually free of sea ice.

Even though the weather was a bit warmer than usual, sailing open vessels across the North Atlantic in medieval times was still a grueling and risky endeavor. Only fourteen of the original boatloads survived the stormy crossing to plant two settlements on the southern tip of what is now known to be the world's largest island. They tilled the reluctant dirt as best they could and sat out the long dark winters in low stone-and-turf houses that were warmed by the radiant heat of their own bodies and those of their livestock as much as by smoky fires. They had the place pretty much to themselves because in those days most native Greenlanders lived much farther up the frigid coast, where the floating ice was more reliable and the seals and walrus more abundant. Rarely interacting with the southern farmers, the native hunters had only colonized the place from more northerly regions of the Canadian Arctic about two hundred years before Eirik showed up.

After 1408 AD, all documented Norse contact with Greenland ended, and extinction of the colonies was complete by the end of the fifteenth century. Nobody knows for sure what did them in. Some scholars blame hostilities with the native hunters. Archaeological studies also suggest that the white settlers overused the land, ruining its tenuous productivity through soil erosion and deforestation. But another compelling possibility is that they succumbed because the climate changed around them.

Floating pack ice had already begun to spread farther south again as early as the fourteenth century. As the global cooling period known as the Little Ice Age gathered strength, warm summer weather became less and less dependable in agricultural Europe, glaciers in the Alps slouched downhill to crush Swiss farms and villages, and the first winter "frost fairs" were held on the consistently frozen Thames River. Because Greenlandic colonists depended on Iceland and Norway to keep them supplied with iron, tools, extra food, and long wooden planks for the construction of vital watercraft, the effects of encroaching sea ice on trade could have been deadly, as could a shortening of growing seasons on already marginal farms. On top of the real and possibly fatal threat that it posed for them, the relentless cooling must have resonated cruelly with the Norse belief that Ragnarøk, the fated end of the world, would begin with a fierce and prolonged "winter of winters."

Whether or not the Medieval Warm Period and subsequent Little

Ice Age really orchestrated the birth and death of the colonies, those climatic disturbances revolutionized the environment of Greenland over the course of centuries and made life first easier and then dramatically harsher for those who lived there. And as the twenty-first century once again leads us into warmer conditions, Greenland has retaken center stage in a new saga of climate change, one whose telling draws a global audience.

Today, Greenland is still shared by Arctic natives and Scandinavians: the modern-day Inuit and the Danes. Tiny Denmark, a diminutive mitten of land and associated islands that constrict the mouth of the Baltic, gained title to that territory in 1814 when the former owner, Norway, lost it in the Treaty of Kiel. Disputes with Norway and the United States cropped up occasionally after that but Danish claims were finally settled in 1933 by an international court. Greenland is now a self-governing administrative division of Denmark, not exactly a colony any more but not fully independent, either.

For the rest of the world, Greenland's coastal rind of habitable land has had only minimal impacts on media, economies, and politics. But Anthropocene warming is changing that quite dramatically.

The melting of Greenland stands front and center in public discussions of global climate change nowadays, and with good reason. The unstable West Antarctic ice sheet is also on the radar screen, but Greenland is different. For one thing, it is home to more than fifty thousand people; there are no "Antarcticans" in the Southern Hemisphere. The stability of the Antarctic ice is more significantly compromised by the breakup of floating ice shelves that are now becoming less and less capable of holding the seaward ends of land-based glaciers in place. And although the northern ice mass is less likely to slide bodily into the ocean than the West Antarctic sheet seems to be, much of it lies far enough from the pole to thaw extensively in summer. The southernmost tip rests well below the Arctic Circle, at about the same latitude as Anchorage, Oslo, and Helsinki. Greenland is a glacial dinosaur, a holdover from the last ice age that makes its own self-refrigerating weather. The vast expanse of reflective whiteness repels enough solar energy to keep it cooler than it would otherwise be, and the great height of the central ice domes—as much as 2 miles (3.4 km) in places—adds to the effect. Mean annual temperatures in the highest northern sites are fiercely cold, close to 23 degrees below zero (-30°C).

But at lower elevations and in places exposed to milder conditions near the coasts, things are changing rapidly. Tourists, journalists, scientists, and local residents now tell of meltwater lakes filling and draining from furrowed ice surfaces, of glacial blue-and-white avalanches thundering away from the snouts of retreating ice streams, of seal hunters who no longer trust the thinning sea ice on local fjords to support them.

These are the basics of a well-known tale, but they also raise questions that are often left unanswered. How quickly could so much ice disappear? And what might the place be like after the most dramatic changes have run their course and less icy conditions become the norm?

In order to determine how long Greenland's ice cover may survive long-term warming we must know how much frozen water there is to liquefy and how much is lost from year to year. High-tech surveys have recently updated the volume of Greenlandic ice from 624,000 cubic miles (2.6 million km^3) to 696,000 cubic miles (2.9 million km^3), enough to build a symmetrical cube 90 miles (142 km) tall. Space-based measurements, particularly those made by a pair of U.S.-German satellites (GRACE, or Gravity Recovery and Climate Experiment) also reveal more ice loss along the margins than we knew about just a few years ago, and each new report from the field seems to indicate faster rates. Glaciologist Konrad Steffen, of the University of Colorado in Boulder, recently summed up some of the latest findings for me: "GRACE results reveal an ice loss of about 200 gigatons per year now, of which 40 to 50 percent is due to surface melt and the rest to increased iceberg calving. Greenland's mass balance has been negative since 1995, and the rate is increasing." One research group headed by Scott Luthcke, of NASA Goddard Space Flight Center, has equated the annual melting with the flow of six Colorado rivers, and another group estimates that recently accelerated outflows have released enough water to fill Lake Erie. Implicit in such descriptions is the suggestion that the ice sheet is rapidly disintegrating before our eyes. But is it?

To begin to clarify what's going on in Greenland during this early warming phase of the Anthropocene, it helps to maintain a proper sense of scale. An update on how much ice was lost this year is of little use if we don't also know how it compares to the overall ice budget. Reports of Colorado Rivers gushing from Greenland's white flanks make it sound like a dying whale in some Melville novel. But if we understand how big that ice sheet is, such metaphorical rivers shrink in relation to their monstrous source. Consider a recently reported loss of 48 cubic miles (200 km³) of ice per year between 2003 and 2008. If that rate continues far into the future, then the entire ice sheet could vanish in as little as . . . 14,500 years.

A warmer future might speed things up, of course. Computer modelers have already made numerous estimates of what additional warming could do to Greenland, but their conclusions vary because the models don't yet consistently express the full complexity of ice sheet dynamics. If Arctic climates warm by 11 to 12°F (6 to 6.5°C) and stay that way indefinitely, various models suggest that Greenland might take as many as 20,000 years to melt down or as few as 3,000 years. Some models also calculate that a higher sustained rise of 14°F (8°C) could destroy most of it in 5,000—or in as little as 1,000—years. Such conditions are well within the range of possibility for a runaway greenhouse scenario, but even in the most extreme situations the ice will take far longer to disappear than many of us imagine.

When we focus on the *net* loss of ice by comparing outflows to buildups, we also find that Greenland's current contribution to sea-level rise is smaller than one might expect. For now, most of the melting and surging is happening along the margins, at relatively low elevations where milder temperatures and undercutting by seawater speed those processes along. Estimates of the net annual outflow of meltwater vary, but they have recently ranged between 100 and 300 gigatons and accounted for a tenth to a third of the rate of global sea-level rise.

The volumes of runoff and ice wasting away from Greenland are large, not only because the world is warming up but also because Greenland itself is gigantic. Even though it is classified as an island, it spans about 1,550 miles (2,500 km) from north to south and up to 680 miles (1,100 km) crosswise. It's as large as Saudi Arabia, twice the size of Ontario, three times bigger than Texas, and about four times larger than France. If we treat that ice mass simply as an enormous chunk sitting passively under a heat lamp, then we've got nearly 700,000 cubic miles of it to consume by thawing layer after slushy layer away from the surface. Little wonder, then, that the time frames on which surface melting alone can demolish Greenland's ice cover are counted in thousands of years.

But in reality, it's not simply a matter of dripping ice cubes. Much of the loss is dynamic, meaning that sizable pieces are tumbling into the oceans en masse rather than trickling away as surface runoff. If you want to know why we're still so short on specifics about the future of polar ice sheets, consider the following list of complicating processes that we now know are at work under those snowy white surfaces.

Coastal "outlet glaciers" are rivers of creeping ice that drain the main pile and pour their contents out to sea in the form of icebergs. Sometimes they surge and sometimes they slow, and nobody's yet sure exactly why they do what they do at any given time. For instance, the Helheim glacier on the eastern coast recently doubled its discharge but then slowed down quite a bit. In contrast, the huge Jakobshavn ice stream, which drains the west central sheet, doubled its pace after 1992 and showed no sign of slowing as of 2009. Some say that a floating ice shelf that once resisted it has now vanished and therefore no longer stems the flow. Others say that meltwater and wet muds under the ice lubricate it like a banana peel underfoot.

Another complicating factor is that thick ice develops an odd molecular structure under the tremendous pressure of overlying layers, so it behaves more like pliable putty than the brittle cubes floating in a cocktail glass. This transformation endows ice sheets with the gift of motion; when they grow heavy enough, they ooze sideways under their own weight. That's why the last ice age left so many scratches and grooves in the bedrock of formerly glaciated lands; boulders and pebbles became gouges and chisels in the grip of sliding basal ice. When retreat time finally arrived, the northern ice sheets lost their lateral mobility as they thinned from the top downward. Rather than dragging themselves back toward the pole, the stagnant southern margins simply dropped their embedded stone-working tools and decayed in place like snowdrifts in spring.

If compressed ice is freed of its overburden, as would occur after prolonged surface melting, it reverts to its preferred, more voluminous crystal structure. Researchers who drill long ice cores from thick polar ice sheets may wait for weeks while their prizes swell and stabilize in cold storage; otherwise, they can shatter when sampled. In the field, the relaxation and cracking of decompressed layers trigger icequakes and might help to fracture some of the remaining ice. Even more powerful are the massive tremors that shudder through advancing outlet glaciers as heavy bergs snap off and fall away from their crumbling seaward faces.

Such fractures may cause even more complications in addition to direct structural weakening. Meltwater lakes and streams that form on top of the ice in summer can plunge into fissures rather than staying put until winter freeze-up or just running off the edges. The weight of water pounding down into a deep, V-shaped seam can split glacial ice like a hammer-driven wedge splits firewood. And where the fissures meet bedrock, the tunneling meltwater can thaw and lubricate the base of the ice sheet, helping it to slide seaward all the more rapidly.

Radar imagery shows that extensive pools of meltwater lie beneath the Greenland ice sheet, probably as a result of surface seepage and basal melting against the warmer basement rock. If the lake-sized pools become large enough, or if seawater leaks in around the edges and seeps into deep valleys in the underlying bedrock, the overlying sections of the ice sheet might destabilize. I asked glaciologist Gordon Hamilton about this during a conference at the University of Maine's Climate Change Institute. Could such a thick, heavy pile of ice actually float?

"Sure it could," he replied. "Water doesn't compress, so if it seeps down under there it can lift the ice away from the bedrock." I recalled a presentation earlier that morning that showed a snakelike coastal glacier rising and falling slightly with the tides, almost as if it were breathing. "And if the ice pulls back from the rim of a rocky basin that it sits in, seawater might fill that space, too. Tidal cycles could then warp the overlying ice up and down until it cracks off."

Still another way to help demolish an ice sheet is to lower its high, cold surface. At present, Greenland counteracts some of its marginal ice loss by piling more snow on top of itself, especially where the central plateau is so high and cold that it doesn't thaw much. But that's not enough to cancel today's net deficit, which is only likely to grow larger as northern air and oceans continue to warm.

Hamilton continued. "If the ice margins lose lots of mass to melting and dynamic thinning, then the high central portion of the ice sheet will begin to sink lower into warmer elevations where it could start to melt in summer. That melting would drop it even lower, and so on." This process would presumably continue until the pile becomes so thin that it doesn't flow laterally anymore, at which point surface melting would be left to finish the job.

And then there's crustal rebound. A heavy ice sheet bends Earth's crust beneath it like a spongy seat cushion. Remove or lighten the load, and the crust or cushion bounces back. Rebound earthquakes still shake the northeastern United States even though the last ice age ended more than 10,000 years ago, and newly generated rebound is measured routinely in Greenland today. Between 2001 and 2006, the Helheim glacier alone drained enough mass from the southeastern sector to let the underlying bedrock rise by more than 3 inches (8 cm). As Greenland loses more weight, rebound movements might further fracture the remaining ice.

This list of ice-eating mechanisms suggests that recent estimates of Greenland's disintegration rate may have been too conservative. Some glaciologists now worry that a wholesale collapse could follow a perfect storm of surface melting, basal lubrication, coastal surging, and the like. Others, however, strike a more measured tone, suggesting that the drawdown could slow significantly once the main body of ice retreats far enough from the ocean's edge to reduce the thawing influence of marine currents and to stem the outflow of icebergs. But the lack of consensus on such details shouldn't be surprising. We're still new at this, and we don't yet fully understand all the mechanisms at work. As a result, we still don't have many firm answers about the future of Greenland other than the obvious conclusion that continued warming will cause continued shrinkage at any one of a wide range of possible rates.

Geological records of past warmings don't tell us as much about ice stability as we might like, either. What long-term history we do have, however, shows us how stubbornly stable Greenland's ice heap is and tends to favor the less apocalyptic arguments. Technically speaking, it can be rather surprising that the ice sheet exists today at all. It lies at lower, warmer latitudes than larger ice age behemoths that once lay closer to the pole but still disintegrated under the thermal assault of postglacial warming thousands of years ago.

Such resistance to warming apparently also existed in the more distant past, as well. Roughly 13,000 years of warmth during the Eemian interglacial still left between a third and a half of Greenland's ice cover in place. That's amazing, considering that a sediment record from Baffin Island, based upon ancient insect remains, showed that summer temperatures there were an impressive 9 to 18°F (5 to 10°C) higher than today. And an earlier, 30,000-year interglacial didn't de-ice Greenland completely, either.

On the other hand, history also shows that megamelting can and does happen, especially when Earth is more generously adorned with mobile ice than it is now. About 14,500 years ago, a melt volume equivalent to two or three liquefied Greenland ice sheets poured into the oceans over the course of 500 years or so. Because ice cores show that Greenland was still heavily glaciated then, much of the loss must have occurred elsewhere, probably from other ice masses that covered more of the polar regions at the time.

The edges of the gigantic Laurentide ice sheet that once buried most of eastern Canada and the northeastern United States gradually decayed as climates warmed during the early stages of deglaciation, but gigantic sections of the central dome also destabilized suddenly from time to time. One notable collapse occurred over what is now Hudson Bay, perhaps because seawater seeped into a depression under the creeping belly of the ice. The resultant rush of bergs into the North Atlantic disemboweled the Laurentide monster, stranding isolated remnants on Baffin Island, Labrador, and northern Quebec.

Three important points arise from these findings. First, a composite ice volume larger than Greenland's was once destroyed within a matter of centuries. Second, most of Greenland's ice nonetheless survived millennial-scale warmings of the past. And third, dynamic undercutting by seawater probably heaved more polar ice into the ocean than atmospheric warmth alone did. In other words, a fully glaciated Greenland today is unusual not because its thick white mantle should necessarily have vanished long ago; its interior may be too isolated from marine attack to suffer the fate of the Laurentide ice sheet. Rather, it's unusual because it's so alone. If it weren't for postglacial sea-level rise and the low-lying nature of the lands around Hudson Bay, there might still be a second major ice mass perched atop northeastern Canada, too.

The points mentioned above are not necessarily as reassuring as they may sound at first, however. True, most of Greenland's ice survived the Eemian, but the last interglacials were not as hot as the deep future may be, and partial but still substantial ice collapses did happen during those earlier warmings. Above all, the crucial factor that will determine the fate of polar ice is not so much the intensity of future warming as its longevity. Even a moderate 1,000-Gton emissions scenario means tens of thousands of years of higher than modern temperatures, and today's summers are already removing more ice from Greenland than winter delivers to it. When we think in Anthropocene-scale terms it seems clear that most or all of that ancient ice is doomed.

In 2005, glaciologist Richard Alley joined forces with European and American colleagues to publish an unusually detailed computer-generated outline of such a retreat. In their simulations, holding CO_2 levels steady at 550 ppm warms the air over Greenland by another 6°F (3.5°C). That bites a deep notch into the southwestern sector of the ice sheet by 3000 AD, and the circumference of the sheet contracts slightly. In 5000 AD, about a third of the ice remains. More prolonged 550 ppm conditions eventually melt the whole thing.

The most extreme situation modeled by Alley's group involved CO_2 concentrations of 1,000 ppm and about 13°F (7°C) of warming relative to today, a situation significantly less intense than what a 5,000-Gton emissions scenario might be like. In that case, about half of the ice melts down by 3000 AD, and by 5000 AD only a few glacial shreds linger in a rugged range of mountains along the eastern coast.

That study, however, did not fully consider all the dynamic processes that could accelerate the languid pace of simple surface melting, so perhaps the actual de-icing time could shrink to a millennium or so. For now, at least, we should trust the proposed sequence of these events more than their timing.

What might Greenland look like several centuries from today, say, in 2500 AD? Alley's work suggests that a moderate-emissions scenario won't bring much dramatically visible change by then other than some thinning along the coasts. In an extreme case, though, much of the southwestern third of the island could be exposed by surface melting alone.

Now let's imagine that Denmark still exists as a sovereign nation and continues to retain strong political and economic ties to Greenland well into the future. As Greenland emerges from beneath its icy mantle, low-lying Denmark could gain ground overseas while losing out to sea-level rise at home. Removing a third of the ice coverage by 2500 AD in a relatively extreme emissions scenario (or several centuries later in a moderate one) could expose 280,000 square miles of new land (725,000 km²). Denmark, in contrast, covers only a little over 16,600 square miles (43,000 km²).

The newly unveiled landscape will surely support some kind of vegetation, and Greenland will begin to deserve its name. What could grow in such a place? We're talking Arctic Circle here, so the mercury will still drop substantially in the darkness of winter. If it warms by several degrees, then we might expect some mix of tundra and forest to develop wherever suitable soil accumulates along the rocky coastlines. But we have better ways of making such guesses.

Records that were kept by Norse settlers during the Medieval Warm Period describe birch woods with trees up to 20 feet (6 m) tall and hillsides thick with hay and willows. Scrubby birches and willows still grow in coastal Greenland today, and much of the rugged coastal landscape already turns green in summer, if only with soft mosses, tufted cotton grass, alpine flowers, and scruffy shrubs.

Geohistorical evidence also provides glimpses of what grew there during longer warmings of the past. Marine sediment cores collected near Greenland's southern coast contain pollen and spores that were blown out to sea during the last million years. During one warm interglacial 400,000 years ago, much of the land was cloaked in conifer forests. And DNA in gritty deposits that came up in the bottom of a mile-long ice core included genetic material from spruce, pine, yew, and alder trees as well as from butterflies, moths, and beetles.

According to some experts, it's not so much Greenland's high latitude that keeps forests at bay today as it is the cold, fierce winds that roar down off the sides of the mountainous ice sheet. Whenever past warmings pushed the margins of the ice far enough away from the sea, trees moved right in. Coastal forests of birch, willow, spruce, pine, yew, and alder, then, are a good bet for 2500 AD, and local citizens of those times will have access to firewood and construction materials that must now be imported at great expense.

We might also expect to find more farmers raising the usual Scandinavian crops, such as potatoes, turnips, sugar beets, cabbage, carrots, and rye. This would generate fresh local produce for folks who may not be able to afford costly imported vegetables. With abundant homegrown silage on hand, it should likewise become easier to raise livestock. Sheep and reindeer are already farmed successfully in southern Greenland, and a recent article in *Spiegel International* reports that the government hopes to launch a dairy industry there as the warming continues.

This new world will awaken while an open-water fishery is booming on the ice-free Arctic Ocean. Greenlandic harbors will be well positioned for fishing and trade with other nations both around and over the pole, and local shipyards and fish-processing plants could thrive. Depending upon what species are available then in the acidified northern ocean, the Greenlanders might harvest cod, halibut, shrimp, and salmon as they do now, but yields will probably rise and southerly species such as mackerel should also lengthen that list. By one estimate, simply establishing a new, self-sustaining cod fishery in western coastal waters could generate as much money as Denmark now donates to the Greenlandic economy annually. In any case, the accessible range of fishing grounds will be hugely expanded during the ice-free summers of the new north.

Fishing might also expand in the interior. A survey of under-ice topography, published in 2001 by British glaciologist Jonathan Bamber and colleagues, shows that central Greenland has been deeply deformed by the weight of its overlying ice. If the southern third of the ice disappears, then the tip of that central depression will be exposed and will trap meltwater. We might therefore expect large, deep lakes to form there. Perhaps they'll be stocked with salmon and trout for recreation and commerce.

As for the underlying rock itself, the Danish geological survey has mapped existing coastal exposures in order to preview what lies farther inland. Geologically speaking, Greenland is an eastern outlier of the Canadian mainland, which is generally known more for its vast granite and gneiss formations than for eye-popping gems and metals. However, some noteworthy impurities sparkle in that otherwise unremarkable matrix.

Today, at least ten gold-bearing localities have been mapped in the ice-free regions, and several molybdenum and lead-zinc mines are currently active. Rubies occasionally turn up here and there, and diamond-laden kimberlites are common along the southwestern coast. A green mineral found near Nuuk bears the unofficial name of "greenlandite" with an eye toward future gem markets. As the ice pulls back, prospectors will find new deposits of such stuff, as well as the copper, platinum, uranium, titanium, nickel, and iron formations that are already known but not yet heavily exploited. And petroleum companies are planning to exploit rich carbon-fuel reserves offshore; according to one assessment by the U.S. Geological Survey, the East Greenland rift basins alone may contain 9 billion barrels of oil and 86 trillion cubic feet of natural gas.

Put all of this together, the open land, the crops, the livestock, the seafood, the minerals, and the fossil fuels, and you have the makings of a prosperous economy—at least, barring major oil spills. But in any environmental reorganization of this magnitude, of course, some must lose while others gain. Traditional Inuit hunting culture, which depends on solid ice for travel and for the environmental needs of seals and other prey, may be lost as the planet continues to warm, and petroleum extraction comes with a risk of major spills. But by most measures, Greenland as a whole will be among the world's overall winners as the Anthropocene proceeds.

What are the greatest uncertainties in this picture? I see two of them: how fast the ice actually melts, and who controls Greenland when the ice goes. It is surely the latter, cultural aspect that will play the more important role in determining how the story of this new world plays out. Will some superpower manufacture an excuse to invade the place? The United States has a military base at the northwestern settlement of Thule (Qaanaaq), after all. Will World War X have left us in a quasi-Neolithic state by then? Or will the native Greenlanders cut their last political ties to Scandinavia and tell the Danes to stay home? Only time will tell.

Now, while we're still on the topic of long-term ice retreat, let's look even farther ahead in time. What might Greenland look like . . . naked?

Leaving unresolved the questions about mechanisms and dates aside, let's imagine how the basic steps of such an unveiling might progress. A study headed by British Meteorological Office scientist Jeff Ridley has modeled a Greenlandic meltdown in detail, and I'll use it here to help sketch a likely sequence of events.

In this scenario of intermediate-scale severity, which is driven by continuous CO_2 concentrations of 1,160 ppm, a widening halo of open ground encircles the shrinking ice sheet by 3000 AD, and the lower third of the island lies fully exposed. But the process slows down after that. The sides of the ice mass have become steeper, so they undergo less direct heating from the sun, and bergs no longer drop into the ocean because the ice margins now lie too far inland; good news for northern shipping. Meltwater lakes, some of them more than 100 miles (160 km) across, ripple under brisk Arctic breezes as they collect runoff from the stubs of ancient ice beside them. Cold-weather farms and forests keep the land green in summer, and that darkening of the landscape launches a new kind of local weather pattern.

Removing much of the high, cool, reflective ice sheet warms Greenland dramatically, and the combination of darkening and lowering of exposed surfaces in the ice-free halo raises local air temperatures by nearly 23°F (13°C) in summer. Now lying close to sea level instead of hundreds or thousands of feet in the air, even the winter landscapes are 13°F (7.5°C) warmer than they would be with the ice still in place. In combination with the greenhouse effect, these changes have boosted Greenland's average annual temperature by 19°F (10.5°C) or more. At Thule/Qaanaaq, for example, this means average July temperatures of 60°F (15°C) at the height of the summer melt season and average midwinter temperatures of -13°F (-25°C).

As summer ripens, warm chimney-plumes of air float above the defrosted regions and then bend sideways over the central ice sheet. From there they cool, sink, and slide back downslope, closing the loop on a new kind of weather system. The streams of dense cold air rushing down the ice sheet's steep flanks cool the marginal areas where most summer melting occurs, thus delaying their final demise.

A matching circulation cell points seaward and drives moisture-rich winds ashore from the North Atlantic. As a result, Greenland's summer weather varies considerably from place to place. Cold glacial winds buffet the ice margins, afternoon sea breezes freshen the coasts, and more rain and snowstorms than usual form over the intermediate uplift zones.

In 4000 AD, as much as a third of the central ice sheet still leans against the eastern mountains. As the sheet sinks lower it presents less and less of a barrier to westerly winds that tend to skirt Greenland's southern tip until they blow straight across the mainland more and more easily. This reshapes the paths of storm tracks over Europe and cools the Barents Sea region that lies east and downwind of Greenland, increasing its floating winter ice cover. If polar bears and ringed seals still exist this far into the future, then the localized cooling here means that Svalbard and Nova Zemlya might be good places to look for them.

Finally, in 5000 AD, Greenland lies nearly ice-free under the midnight suns of summer and the noontime stars of winter. Along the eastern coast, a range of mountains up to 12,000 feet high (3,700 m) cradles the last shrunken glaciers. If Santa Claus still exists in the farther reaches of the Anthropocene, that mountain refuge could be where he makes his last wintery stand; in fact, most Scandinavian children will already tell you that "Nisse" or "Tomte" has been living there all along, rather than at the North Pole where American kids now write to him.

By this time, a prominent new feature has appeared on the national map. Under the thickest portions of ice, the rock was depressed far below sea level. With most of the ice sheet gone, rising seas pushed through narrow northern channels and flooded that central depression. Now a vase-shaped, island-studded fjord 500 miles long (800 km), up to 310 miles wide (500 km), and perhaps 1,500 feet deep (450 m) straddles Greenland's down-warped spine. Spruce and birch forests crowd the shores except where settlements, roads, and farms carve up the greenery. Whether people of those future times ply the waves of this fjord under the power of petroleum, steam, sail, or paddles, ply it they do. It's safely sheltered from the heavy swells of the open ocean, and with outlets to the sea it provides easy access to rich northern fishing grounds as well as inland water routes to much of the country.

When I first learned about this new waterway I realized that it might some day require a name. Why not propose one now? Perhaps some future geographer will find an old copy of this book and put the name on an official map. At first, my mischievous side was tempted to make it unpronounceable by Danes. I once lived in Denmark and found, to my amusement, that many of them struggle to pronounce "refrigerator," just as most Anglophones can't say "røged ørred" without gagging. Fridge-fjord briefly came to mind. But a better choice might be Ny Fjord

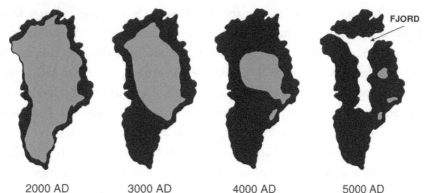

| 2000 AD | 3000 AD | 4000 AD | 5000 AD |

Progressive melting of Greenland's ice cover under an emissions scenario of intermediate intensity. *After Alley et al., 2005*

(NEE fee-YORD, translating to "new fjord") until, more appropriately, some native Greenlander names it.

In any case, this scene won't last forever, and it might not even last through much of the Anthropocene. Ny Fjord will eventually be destroyed by a delayed reaction to the very de-icing process that produced it.

When the last ice age crushed the northern circumpolar lands, it pushed bedrock hundreds of feet down into the viscous mantle below. As the ice retreated, the crust bounced slowly back, and much of it is still rebounding. A map of uplift rates in Scandinavia, for example, reveals a ghostly silhouette of the ice that once squashed it, centering a bull's-eye of concentric rings on the northwestern tip of the Baltic Sea. In that corner of Sweden near Luleå, the land rises a third of an inch (8 mm) every year. Move southward along the Baltic coast and the rates decrease; Stockholm is rebounding half as rapidly, Malmö is a quarter as fast, and so on. Something like this will happen in Greenland, too, once the Anthropocene defrosts it.

In 5000 AD, the unburdened land lifts gently skyward, even though the weight of the water in Ny Fjord resists it. The bottoms of the deepest central basins will already have bounced about a third of the way back as quickly as the ice left them, but the rest of the recovery is sluggish. In 5000 AD, the floor of Ny Fjord rises by 2 or 3 inches (6 to 7 cm) per year, and chronic earthquakes rattle the towns and farmsteads of central Greenland.

By 8000 AD the fjord is about half as deep as it was when it first formed. The coastal channels are still open because the coasts haven't risen as much as the central depression has, and global sea level is much higher now that a great deal of polar ice has melted. But eventually, tens of millennia farther down the long tail of the CO_2 curve, rebound leaves the outlets high and dry, cutting Ny Fjord off from the Atlantic. Freshwater runoff dilutes the marine brine, and the fjord becomes a lake.

This sort of thing happened in eastern North America shortly after the last ice age. The newly deglaciated Saint Lawrence River corridor still lay so far below sea level that saltwater rolled in from the east and flooded the Champlain Valley when the ice left. For several centuries, seals and whales swam within sight of what is now the waterfront district of Burlington, Vermont, and fresh-looking shells of blue mussels and white *Macoma* clams still turn up in gravel pits near Plattsburgh, New York. As the land rebounded, the short-lived Champlain Sea lost its lifeline to the ocean, stranding any seals, whales, mussels, and clams unlucky enough to be born into those transitional times. Gradually, the water body was diluted by river and rainfall inputs and became what we now call Lake Champlain.

In 10,000 AD, Ny Fjord Lake might still be larger than the state of Florida, but its rebounding floor is much higher now, and the water is only about 300 feet deep (100 m). As more millennia pass, bathtub rings of raised beach deposits encircle the contracting shorelines, some of them dotted with the remnants of former harbors and the crumbling foundations of former port towns.

Climate whiplash is long past and a slow cooling trend is lowering temperatures ever closer to those of today. The change is too subtle for the average Greenlander to notice in the midst of natural year-to-year climate variations, just a small fraction of a degree per century, but the gradual ocean retreat that accompanies it is more readily noticeable in a nation that makes much of its living from the sea. Most who live and work on the ocean's edge then consider the retreat to be a simple fact of life and easily adapt when necessary, as water-folk did during the more rapid changes of the twentieth and twenty-first centuries. Even so, long-term cooling makes each successive generation of Greenlanders expend a little bit more energy in order to maintain the same levels of indoor comfort that their parents enjoyed. Growing seasons cool and shorten over the course of centuries, and winter ice grips coastal waterways more and more tenaciously in spring.

For some people in that far future who have enough of a sense of history to realize what's happening, these inexorable changes may be a source of unease. Others revel in the dramatic. If an active media industry still exists in 10,000 AD, headlines may warn that "Global Cooling Threatens Greenland with Icy Doom." Newscasters may try to boost ratings by describing in lurid detail how colder weather increases the risk of death by exposure and ice-related injuries, and the term "carbon crisis" might be used in reference to a shortage, rather than an excess, of greenhouse gases. Some who notice the changes might yearn for the slightly warmer, greener, gentler settings that their grandparents grew up in, as Norse homesteaders probably did while encroaching sea ice slowly locked them away from the rest of the world at the close of the temporary warm period that had brought Eirik there centuries earlier.

And some day much farther down the carbon curve, the rebounding crust that once cradled the shrinking waters of Ny Fjord Lake will rise high enough to tip the last of them into the Atlantic. One can only guess when that might be; perhaps in 50,000 AD. Greenland's vertical profile will have oscillated from a dome of ice to a sunken bowl of water to a rising dome of stone. The forested, farmed, and settled landscape will slope upward from west to east, and rivers will rush seaward down formerly ice-bound valleys to drain a rugged spine of mountains on the eastern rim of the great island. It is those mountainous highlands that will eventually spawn a new, ground-crushing ice encroachment when our carbon legacy finally fades to insignificance.

Greenland probably wins more than it loses in this long view of the future, even though it can't keep Ny Fjord in the end. Meanwhile, tiny Denmark shrinks as the sea rises. By the time Greenland loses all of its ice in an extreme-emissions scenario, additional losses from Antarctica could contribute to a total sea-level rise of 40 feet (12 m) or more, enough to submerge the large, flat Danish island of Lolland and nearly sever the Jutland peninsula from mainland Europe. With that future in mind, it's all the more interesting to watch Denmark now loosen its former grip on Greenland, one of the few places that's likely to benefit substantially from global warming. By doing so, it might be letting go of a future lifeline.

When Greenland gained home rule within the Danish Common-wealth in 1979, the Home Rule Parliament left Denmark in charge of foreign policy, defense, currency issues, and the judicial system. Pressure is now growing amid Greenland's mix of 50,000 Inuit and 6,000 Danes to further increase their national autonomy, and in November 2008, local voters passed a nonbinding referendum on full independence. From their local perspective, Anthropocene warming may increasingly favor complete separation, especially once the Arctic Ocean becomes ice-free, more vegetation begins to thrive, and the shrinking ice sheet exposes more and more of Greenland's hidden treasures.

With Denmark's long-term survival at stake, the Danes will have strong incentives to try to convince their old friends to remain within the fold or, at least, to remain friends. Sooner or later, they're going to need Greenland much more than Greenland needs them.

10

What About the Tropics?

*Essentially, all models are wrong, but some
are useful.*   —*George Box, statistician*

Pale wisps of white quartz sand slither and hiss as a hot wind sweeps
them across the hard-packed floor of what was once Ferguson's Gulf. The
broad, shallow bay had dimpled the western shore of Lake Turkana, an
emerald-tinted sea of alkaline water nearly 180 miles (290 km) long that
glistens in the harsh, nearly roadless desert of northern Kenya. A sun-
roasted spit of fine sand and coarse, spiny scrub circles the desiccated bay
like an arm reaching out from the mainland to cradle a basket. A basket
that once held fish, and lots of them.

On this day in 1988, Ferguson's Gulf holds little more than camel
droppings and the sandal prints of Turkana nomads. A jetty that once
stretched hundreds of yards from the gently sloping shore to reach boat-
friendly depths is now a skeleton of rusted iron pilings, its boards long
since scavenged and burned for firewood. A short distance farther inland,
a large building that resembles an airplane hangar stands empty, the wind
moaning through holes in the sheet metal shell. During the late 1970s
Norway spent millions of dollars to build this fish refrigeration and pro-
cessing plant, hoping that it would help to coax skeptical herders of cattle
and goats to settle and net the huge Nile perch and other fish that spawned
in the jade green shallows of the lake. The aim was to lift the Turkana
people out of poverty and drought sensitivity into the supposedly stable
money economy of modern times.

Very quickly, things went wrong. The cost of cooling the enormous building in the fierce desert heat outweighed any profits the filets

provided. And the reluctant Turkana, who had traditionally shunned fish as a starvation food, owned no boats for deep-water fishing. Instead, they harvested the gulf as if it were an aquaculture pond. Its flat bottom let them wade neck deep in the murky waters, towing handheld nets and baited lines behind them. Crocodiles pursued the same fish in the same place, often cruising uncomfortably close to the fishermen. Sometimes, unwary children became their prey instead. Meanwhile, newly sedentary livestock nibbled the already sparse, mostly thorny vegetation bare in an ever-widening ring around the settlement, further locking their owners into an increasingly tenuous, lake-based lifestyle.

And then the lake level began to fall. Ferguson's Gulf shrank, then fused with the surrounding sands, leaving the Turkana bereft of their primary fishing grounds. Many headed back into the hills to escape the overgrazed zone with what remained of their herds. Apparently, the project planning team forgot or failed to recognize that lake levels can change, and that Lake Turkana's surface has a long history of doing so. An expert who was later sent from Norway to evaluate the defunct fish-factory-in-the-desert ruefully summed it up with the question: "How could we have been so stupid?"

Jetty leading into the center of Ferguson's Gulf, Lake Turkana, Kenya, before and after the lake level fell; top photo taken in 1981, bottom photo taken in 1988. *Curt Stager*

In hindsight, it can be easy to dismiss this story as just another interventionist boondoggle, and it seemed like one to me as I drove across Ferguson's Gulf seven years after boating on it in 1981 and having seen the short-lived fishery while it was still in operation. But such mistakes are in fact easy to make. The trap is readily reset when well-meaning people try to help others without fully understanding their physical and cultural settings. Nowadays, with threats of impending climate change urging people to reach into their wallets or policy manuals on behalf of developing tropical nations, it's all too easy to overlook one troublesome fact. Tropical climates are among the least understood of any on the planet. How can we prepare effectively for coming changes if we don't know exactly what is going to happen?

In the latest IPCC assessment report, the tropics received less attention than higher latitudes and the world as a whole in the overviews and predictions for the twenty-first century. That was not because the tropics are unimportant; they cover two-fifths of Earth's surface and support roughly half of all humanity. Some say that this kind of imbalance occurs because most climate research originates in extratropical nations, and it tends to focus more often on northern winter and summer conditions than on the "off-season" months when most equatorial rains fall. Others add that precipitation patterns are generally more important in the tropics than temperature variations are, but rainfall can be difficult for computers to model and forecast accurately. And most experts agree that we have less background information to work with when we study tropical climates than when we consider those in North America or Europe. The United States, for example, has a well-funded, frequently updated, and easily accessible online repository of high-quality weather station data called the U.S. Historical Climatology Network (USHCN). No such network exists for most of the underserved and underrepresented tropics.

For some of us it may seem odd to imagine the greenhouse effect posing a threat to places that are already hot. But if we're to understand climate change on a global scale, we have to know what is going on at the planet's midriff as well as in its cooler places. To do otherwise is to risk repeating the errors that are humorously but effectively described in an old Indian folk tale that I enjoyed as a child. In that story, each of several blind men briefly touched different parts of a large elephant and then offered a seemingly reasonable but wildly inaccurate assessment of what the beast actually was. The man who touched the trunk concluded that the elephant was like a snake, the one who felt a massive rough-skinned leg called it a tree, another man who leaned against the beast's broad flank thought it was a wall, and so on. In hopes of sketching a more complete picture of global climate change than that misguided group might produce, I'll focus mainly on Africa and Peru where most of my research has been centered, while keeping in mind that much remains to be learned about how tropical weather works and what it will be like in the future.

Consider the situation for tropical Africa, which is typical of most other low-latitude regions, as well. The 2001 IPCC assessment reported that "deriving regional climate change predictions [is currently] impossible," and "the potential effect of climate change on drought in Africa is uncertain." The 2007 IPCC report cautioned against "the over-interpretation of results, owing to the limitations of some of the projections and models used," and said that although droughts have long ravaged northern and eastern Africa, "it has not been demonstrated that these droughts can be simulated with coupled ocean-atmosphere models." Tropical climatologist Richard Washington was recently quoted in *Nature* as saying, "You can get any result you like" with an African climate model. And an exhaustive review of regional simulations posted online by the Royal Netherlands Meteorological Institute (KNMI) shows a confusingly diverse range of possible hydrological futures for various subsections of the continent.

Because of this complexity and uncertainty, much of what you may read and hear about the future of tropical climate is questionable. In 2006, for example, an article in *The Independent* said that Africa will warm faster than the rest of the world and will face "the greatest catastrophe in human history," citing economist Sir Nick Stern and various computer-model projections of West African droughts to support the claim. In contrast, a recent scientific report entitled *The Copenhagen Diagnosis* used similar models to show, as most sources do, that it is the poles, not the tropics that are warming fastest, and they also concluded that wetter conditions may occur in the West African Sahel, representing "a rare example of a positive tipping point" in response to global warming. Such contradictory predictions, along with the widely divergent model readouts on the KNMI website, suggest that deciding which vision of the near future to trust is more of a gamble than one might wish. And the same situation exists in other tropical regions as well; according to geophysicist Venkatachalam Ramaswamy in a recent issue of *Eos*, the climatology of Asia remains troubled by "insufficient understanding of intermodel differences" and "inadequacy of our climate models to resolve [regional] spatial scales." Fortunately, sound scientific footing in that regard does exist, and much can simply be deduced from the basic nature of temperature and precipitation in the tropics.

Temperatures are already rising in the lower latitudes, though generally less so than in the polar regions where an extra thermal boost from the loss of reflective snow and ice is currently making them heat up faster than the global average. But people have lived longer in the warm tropics than anywhere else on Earth; we evolved there, after all. Life has survived abundantly in hot places for hundreds of millions of years, and it still thrives there today more than anywhere else. If you want to see thousands of species of frogs, fishes, flowers, and fungi, you go to the Amazon, not the Yukon.

In this context, it may seem odd that most of the imagery that we associate with global warming makes hot climates seem frightening, even deadly. Images of planets on fire or sweaty, red-faced Earths with thermometers sticking out of their mouths are good for gaining attention or raising money but less so for supporting well-informed discussions. Warming is indeed a problem, especially to organisms and societies that are adapted to stable or cooler conditions, but high temperatures themselves are not necessarily a deadly threat to those for whom they are already normal.

As the world warms, most tropical locales will not face the physical tipping point that amplifies changes at high latitudes and altitudes: namely, the melting temperature of ice. In places where things now freeze on a regular basis, future heating may have abrupt and dramatic physical effects from the collapse of ice sheets to the unsheathing of polar seas. But in already hot places there is no obvious ecological boundary to cross on a rising temperature curve. In the Arctic, it warms and warms and then suddenly there's no ice for walruses to haul out on between dives, and the shift from snow cover to heat-absorbent soils, rocks, and vegetation drives temperatures radically higher. When the weather heats up further in most of the tropics, it mainly just feels that much warmer.

I'm sure that most residents of hot places would prefer not to see temperatures rise further. But as we weigh the human costs and benefits of a changing future, it's worth remembering how *Homo sapiens* copes with heat in more realistic terms than those we often encounter in the mainstream media.

I'll never forget gaping in amazement as columns of muscular French Foreign Legionnaires jogged and maneuvered amid the rippling mirages of Djibouti, a furnacelike pocket of lava ridges and troughs that notches the Horn of Africa between Ethiopia and Somalia. The hammering sun and thick, humid air intimidated me as a newcomer, and shortly before I arrived, a carload of French tourists died of heat stroke within hours of a breakdown on a remote desert road. However, off-duty legionnaires later told me that they eventually acclimatize enough to do what has to be done outdoors, regardless of the heat. And after hours, the streets of the capital are alive with off-duty soldiers and local folks because it becomes quite a bit cooler there after sundown. Even in the torrid outback of Djibouti, people love the land that they were born into, and generations of pastoral nomads have vigorously defended their claims to that harshly beautiful landscape from others who have sought to take it for themselves.

These examples show that high temperatures are not necessarily intolerable, especially if one is used to them and not also suffering from extreme poverty, disease, or war on top of everything else. It is mainly change itself that threatens life as we know it more than any particular thermal set point, and the global cooling that eventually follows the future climate whiplash period will also cause problems for many of our descendants, just as the opposite trends do now.

Misunderstandings of tropical climate history are also common. One of the most surprising geological discoveries of the nineteenth century was the realization that vast ice sheets have repeatedly covered much of Eurasia and North America during the last 2 to 3 million years. With the legacy of those past coolings in mind, many scientists assumed from the shortage of glacial landforms in the tropics that low-latitude climates have always been stable, and that the high biodiversity of the tropics reflects a benign climatic setting. But those assumptions are wrong.

What "temperate zoners" often fail to realize is that climate involves more than just temperature alone. I grew up in New England believing that seasons are either cold or warm, and that major climatic shifts of the past centered on the advance or retreat of ice sheets. But if I had grown up in Nairobi or Jakarta, the main seasons would be either wet or dry, and the most noteworthy weather-related disturbances in local history would have involved flooding or drought. Precipitation is just as much a factor in global climate change as temperature is, and in most of the tropics it is by far the more important of the two. Unfortunately, it is also more difficult to model and predict.

Scientists have only known much about the central roles of rainfall and drought in the long-term ecological history of the tropics for the last half century or so, since paleoecologists began to analyze sediment cores from tropical sites. Among the first was my graduate advisor at Duke University, Dan Livingstone, who traveled through much of East Africa with his graduate students during the 1960s in order to study vegetation and lake levels of the past as a way of deciphering the nature and causes of tropical climate variability.

Using now-classic techniques, Livingstone and his students drove long metal pipes into the soft sediments beneath lakes such as Victoria, Tanganyika, and Cheshi (where a crocodile sank their boat but let them escape). Back in the lab, they studied ancient pollen grains and the glassy shells of diatom algae under the microscope and reconstructed, layer by sedimentary layer, a record of African climate that stretched thousands of years into the past. What they found surprised the scientific community. Not only had ice ages brought climatic changes to the tropics after all; the paleoclimate records showed that those changes were both extreme and quite different from those at high latitudes.

When the north grew colder and snowier East Africa cooled somewhat as well, but instead of becoming wetter, it parched. Near the end of the last ice age, East Africa became so dry that local forests gave way to dry savannas. About 17,000 to 16,000 years ago, the surface of 4,800-foot-deep (1,470 m) Lake Tanganyika sank low within its Rift Valley basin, and Victoria, now the world's largest tropical lake, disappeared altogether. Elsewhere in the presently wet monsoonal regions of southern Asia, geohistorical investigations reveal similarly intense drying, as well. The changes made physical sense, after all; warmth drives humid air upward where it condenses into rain clouds, and cooling tends to reverse the pattern. Paleoclimate records have repeatedly shown that this basic rule of thumb—warming means wetter, cooling means drier—applies to long-term climatic history in much of the inner tropics. This, in turn, eliminates environmental stability as a possible cause for high species diversity in the tropics, and the mystery still remains unsolved in scientific circles.

We have fewer geological records of interglacial warm periods from the tropics than we do for the temperate zone, but most do indicate wetter conditions during those hotter times. Sediment records from the eastern Mediterranean Sea show that the Nile River poured huge floods of buoyant freshwater atop the salty waves during the Eemian and again during the early Holocene. During that last warm spell, the Sahara was a game-rich savanna and Lake Chad, now a relatively shrunken puddle of brine, was a huge inland sea of freshwater. Cores from submerged deposits off the coast of Mauritania clearly document the end of that warm Holocene wet phase; about 5,000 years ago, easterly winds began to sweep more and more desert dust into the ocean there, signaling the desiccation of northern Africa.

More recent work, however, has revealed some striking exceptions to the warm-wet, cool-dry rule. My own research (funded by American readers' tax dollars, thank you, through the National Science Foundation) has shown that although Lake Victoria shrank during major coolings of the distant past, it rose rather than fell during the chilliest phases of the so-called Little Ice Age 600 to 200 years ago. Although most African lakes did not follow that unusual pattern, it has also been recorded in sediment records from Lakes Naivasha and Challa in Kenya and Tanzania, respectively, and it warns us that tropical rainfall conditions can vary in unexpected ways, both geographically and over time.

Does this review of history tell us anything useful about the climatic future of the tropics? Yes, but not as much as we would like. Whatever Anthropocene warming may do to low-latitude climates in the future, its most important effects will probably involve precipitation more than temperature. But recall that the warm-wet, cool-dry pattern doesn't always hold true, and not all regions respond to environmental changes in the same way. Furthermore, many of those long-term shifts of the past resulted from natural cycles in the orbit and orientation of Earth, which were fundamentally different from the intensification of the greenhouse effect that we're currently causing. Nonetheless, we can be fairly certain that many future changes will be linked to the primary weather system of the lower latitudes, a meandering belt of clouds and rainstorms that climatologists call the intertropical convergence zone, or ITCZ.

You can clearly see its trademark—or rather, its watermark—in satellite images of Earth. All along the equator, the land turns green wherever the ITCZ makes regular visits during annual rainy seasons. This is where you find the great rain forests and rivers of the Amazon, Congo, and monsoonal Asia and Australia. Farther poleward along either side of those equatorial green belts the colors fade into what seem, by comparison, to be dust-colored dead zones. These are among the world's largest dry places; the Sahara and Namib and Kalahari deserts, the Arabian Peninsula, and the brick red outback of Australia.

The ITCZ runs on solar energy, and some call the massive atmospheric machinery that drives it the climatic "heat engine" of the planet. The mechanistic metaphor is appropriate because enormous rising and falling loops of air—called Hadley cells—spin like twin sets of interlocking cogwheels as the sun's most direct rays cook the tropics. Intense solar heating warms and expands the overlying air so it floats upward in the churning central zone between the cogs, the ITCZ proper. There it condenses into clouds and spreads against the base of the stratosphere like smoke against a ceiling. Then the cooled air sinks back down at higher, drier latitudes from whence it pushes ground-hugging trade winds back to the central rising zone. That cyclic motion links convective rains, cloudless deserts, and reliable sailor-friendly winds into a single cohesive system that has dominated tropical climates through the ages.

The two tropical boundary lines rest at 23.5 latitudinal degrees north and south of the equator. Every June the North Pole leans most steeply inward at the sun, and that makes the most direct, high-energy light strike the Tropic of Cancer near the latitude of Calcutta, Aswan, and Havana. Stand anywhere along that line at midday in late June and the sun will beat straight down on your head. In December, when the Southern Hemisphere's summer begins, a similar leaning of the South Pole makes direct sunlight strike the southerly Tropic of Capricorn near the latitude of Rio, Windhoek, and Alice Springs.

Wherever the zone of vertical rays wanders between those parallel tropic lines, the ITCZ follows. A creamy band of clouds sweeps back and forth with each passing year, and as it passes overhead, tropical residents enjoy their seasonal doses of rain. In this manner, the drumbeats of thunderstorms drive the seasonal rhythms of life in some of the world's most densely populated regions, from the summer monsoon realms of India and Pakistan to the steep hillside farms and forests of Rwanda and Burundi.

If both common sense and many global climate models are correct, then future warming will drive the heat-sensitive Hadley cogwheels more vigorously and spread the inner zone of ITCZ rains over a wider range of latitudes. At the same time, warming oceans will evaporate more water vapor into the system, so more rain could fall from the rising, condensing columns of air within the ITCZ. This is a recipe for more abundant and more extensive precipitation in the tropical monsoon regions, and it helps to explain why most models predict twenty-first-century wetting in southern Asia and most of equatorial Africa and South America.

Sites that lie at higher latitudes near the outer boundaries of the tropics, though, might face the opposite situation. Where sinking air on the descending flanks of the Hadley cells now inhibits cloud formation, some dry regions could become even more arid. To put it simply; global warming is likely to intensify whatever is currently going on in the tropical climate system. But as tropical circulation intensifies, both wet and dry regions may also expand their ranges poleward, and that could lead to both negative and positive changes in the transitional zones. In the Sahara, for example, the most harm to people and ecosystems might occur along its northernmost margin, where fiercer droughts could reduce rainfall in what are now the somewhat less arid, more densely settled locales of the Mediterranean coast. Along the desert's midline and southern edges, however, broadening and intensification of the ITCZ rain belt could make life easier. Latitudinal shifts of this sort were important features of the Eemian and PETM-scale warmings of the distant past and are already being reported for some of the prevailing wind tracks in the tropics and the southern temprate zone, so it would be reasonable to expect more of the same in the Anthropocene future, too.

One of the tropical regions that is most likely to face serious water problems in the near future is the long, narrow, hyperarid coastal strip of land between the eastern Pacific and the rugged crest of the Andes. Just a few inches of rainfall on it in a typical year, and yet more than half of the population of Peru manages to live there.

When I first visited the Peruvian desert in 2006, it was in the company of Dan Sandweiss, an archaeologist at the University of Maine's Climate Change Institute where I hold an adjunct research position. His research had shown that sporadic wet periods strongly influenced local cultures for thousands of years, and on that trip we were traveling north to the Sechura region to study geological evidence of those floods. But even more important to coastal Peruvians than the rain pulses have been the rivers that drain the high peaks farther inland.

As we rode a bus northward along the desolate coast from Lima to Piura, Sandweiss explained what I was seeing in that Mars-like realm of scorched black rocks, rusty gravel, and scalloped, dun-colored dunes. "See the faint greenish tinge on those steep slopes overlooking the Pacific? That's where fog blows ashore and keeps a few plants alive. Everywhere else, it's just barren ground." To me, the scene evoked the earliest days of terrestrial life, when simple vegetation was just beginning to colonize formerly lifeless lands along the edges of the oceans. I wasn't surprised when he added, "NASA research teams sometimes use the Peruvian desert to represent a terrestrial version of Mars."

"Now look up ahead," he continued. The black ribbon before us dipped steeply downward into a narrow, wedge-shaped valley whose wide mouth pressed against the flat blue face of the Pacific. "See how lush it is down there? That's where the Rio Santa flows in from the mountains. You'll see this same pattern repeated from valley to valley all the way up the coast. The rivers keep towns and farms alive here, and it's been this way for centuries. If you want to find archaeological sites, these are the places you want to explore."

As our bus crossed the valley floor, the crumbling remnants of precolonial mounds shared common ground with bustling twenty-first-century settlements. But most surprising to me were the new dry-country farms. For mile upon verdant mile, sprinklers rained precious moisture onto paper-thin veneers of asparagus, artichokes, and other cash crops.

With nothing but desert sands beneath those shallow root systems, the fields were more like hydroponic gardens than dark-earth plots.

"These kinds of farms are appearing all over the place now with the help of new irrigation from the rivers and a recent rise in meltwater flow from the high country. Most of what you see here didn't exist just a couple of years ago. It's a huge boost to Peru's economy, but it all depends on reliable supplies of river water." And that's the worrisome part. As the world warms, glacier-fed rivers will be anything but reliable.

Peruvian glaciers, which represent roughly 70 percent of all tropical ice, are like trust funds. They collect rich windfalls of snow in narrow seasonal windows and then dole some of it back out in steadier dividends throughout the year. In some watersheds, up to half of all stream discharge during the annual dry seasons stems from melting ice and snowfields. In that manner they have maintained lifelines of river water to the arid lowlands for centuries, and roughly three-quarters of Peru's electricity is now generated by hydro dams that depend on such flow. But modern climate change is raiding the principal on those reserves of high-altitude ice.

Many Andean glaciers are now dispensing more water than they take in, and they are visibly shrinking. Total ice cover in Peru has decreased by a quarter since 1930, and some of the smaller glaciers, such as Ecuador's Cotacachi, have vanished altogether. Increasingly worried Peruvians watch local ice streams shrink farther and farther back into the mountains as if they were white fuses sizzling upward. When the last tendrils of ice disappear among the rugged spires of the Andean skyline, an environmental catastrophe could explode downstream.

Life requires water, and precious little of it falls on western Peru, which lies in the rain shadow of the Andes. Peru has faced great difficulties, from the Incan and Spanish conquests of the last millennium to more recent social unrest, and only now have intensive desert agriculture and hydro power emerged to bring hope of more prosperity by harnessing glacier-fed rivers. It seems a cruel twist of fate that those new industries are already under threat from glacial retreat.

As our bus crossed the razor's edge between green fields and barren desert and began to climb back out of the valley, I asked Sandweiss what he foresees happening here. "Well," he began with a rueful look, "there's talk of building reservoirs in the canyons, or even of tunneling through the mountains to channel water in from the wetter Amazon slopes. But this is earthquake territory, so it's not yet clear how to maintain a dam or a water tunnel for very long without putting people downstream at risk."

And what if Peru also becomes even drier as well as warmer? Fortunately, most computer models envision a generally wetter future for the inner tropics as Hadley circulation intensifies. As one might therefore expect, some of the northern mountains are already experiencing a wetting trend, but for some reason, the majority of the Peruvian Andes have been drying in recent years. A regional turnaround to more generous rainfall is predicted by midcentury in most simulations, but the coming decades might still present severe challenges to people and ecosystems in today's currently drying zones.

In July 2009, I returned to Peru in the company of Kurt Rademaker—a Sandweiss graduate student—and four other promising young geoscientists. This time, the focus was Coropuna, a 21,000-foot (6,425-m) lump of ice-caked volcanic rock that towers over a remote highland wilderness of brown, rubble-strewn plateaus and canyons. Several years earlier and just a short distance from our base camp at 14,000 feet, Rademaker had found a beautifully crafted arrow point flaked from pink coastal chalcedony, proving that native peoples had hiked up there thousands of years ago. But what induced them to go to all that trouble?

One would have been hard-pressed to make much of a living on the flanks of Coropuna in the distant past. For starters, one would need to acclimatize; a lungful there yields about half the oxygen that you'd inhale at sea level, judging from the crumpled state of a plastic water bottle that I emptied and closed at our high camp so I could later watch the maritime air pressure flatten it. There are no trees for shelter and little ground cover; the dominant plant there is yareta (*Azorella compacta*), which looks like a green lump of boulder coral. There would be few wild animals to hunt apart from the rock-hopping vizcacha, which resemble long-tailed rabbits, but are more closely related to chinchillas, plus a few sleek and elusive vicuñas (wild cousins of llamas and alpacas). And the local streams freeze nightly despite their close proximity to the equator. All in all, this is a glorious setting for a scientific expedition, but it can also be a challenging place to call home.

A major key to survival in that high country, as in the lowlands, is water, and I was surprised to see so many sky-tinted streams and pools gleaming on the floors of otherwise desolate valleys and depressions, even though the main December-to-March wet season was long since past. Almost invariably, the blue was cupped in moist carpets of green, the unique upland wetlands known as "bofedales." Neither mossy bogs nor grassy marshes, Andean bofedales consist of dense mats of *Distichia,* a strange, low-lying plant whose close-packed, spiky green stems form tough, ruglike versions of the dry-ground yareta. And atop most of those wet bofedales wander the other centerpieces of pastoral life in these highlands—grazing herds of domestic alpacas.

"It's hard to imagine living up here without these bofedales," Kurt mused as we drove a dirt track along the margin of the broad Pocuncho wetland. Alpaca herders in brightly colored native garb waved as we passed. "These people have a tough life up here, but they manage to make a living by selling the wool and meat of their animals in the lowlands. Their whole lifestyle depends on these wet areas, and the first folks

University of Maine graduate student Kurt Rademaker coring a bofedal wetland near Coropuna, Peru. *Curt Stager*

to move up here after the last ice age probably depended on the bofedales as a place to hunt vicuñas." Whether or not the bofedal wetlands actually did exist in those earlier times was what we aimed to discover by coring and radiocarbon dating the local wetland deposits.

As I write these words the results of Rademaker's study are not all in yet, but our preliminary dating of one thick exposure of bofedal peat layers on the flanks of Coropuna shows that the deposits are indeed thousands of years old. In any case, the journey left us with a host of other questions and observations relevant to the story of climate change in the tropics.

Matt Schmitz, an undergraduate student at Pacific Lutheran University who accompanied us to Coropuna, interviewed local herders and found that concerns about climate change have reached even this remote corner of the world. "The people say that they've watched the ice retreat on the mountain," he reported after a long day in the field. "And they're noticing less rain and snow during the wet season and less water in the bofedales." Such observations fit those described in the scientific literature as well; Coropuna has lost a quarter of its glaciated area since 1960 as precipitation in southern Peru has fallen off.

But the retreat of mountain ice, in this case, has little to do with freshets of meltwater pouring off the slopes; few of them actually exist. High ice caps are composite beasts whose upper and lower reaches respond differently to weather, and most of Coropuna's ice and snow lies well above the freeze-thaw elevation where it remains too cold to melt much. Instead, its shrinkage is mainly due to sublimation, the direct escape of frozen water into the air.

You may have encountered this effect yourself if you've ever wondered why the cubes in your ice tray have shrunken so much after prolonged storage in your freezer. This is the central process behind the de-icing of many of Peru's mountaintops and of Africa's Kilimanjaro, too, and it means that drought can be as much of a threat to tropical ice as warming is. Rather than just melting away, much of it is starving for want of replenishing snows. This also means that not all glaciers and snowfields feed rivers significantly. The lower fringes of mountain ice, those that lie low enough to melt in warmer strata of air, do most of the dripping and dribbling that helps to keep rivers running, but they don't represent the main frozen masses on super-high peaks like Coropuna and Kilimanjaro.

Something else must be feeding those lowland waterways. Bofedal wetlands thrive in isolated gullies that lack any obvious link to ice, but water nonetheless pours through them abundantly on its way to the sea. Such wetlands can trap and dispense the groundwater contributions of seasonal rain and snow in much the same way that glaciers do, and bofedales may therefore represent an important but little recognized source of lowland river water in Peru.

That might be good news in terms of future water supplies, except for two things.

First, the Andes are warming. Most computer models still lack the spatial resolution to simulate the climates of diverse, mountainous terrain accurately, but some experts predict that a moderate-emissions path could leave the region 3 to 5°F (2 to 3°C) warmer by the end of this century, and that an extreme 5,000-Gton scenario could boost the rise by 5 to 9°F (3 to 5°C). Whatever the actual magnitude of that change becomes, additional heating is likely to shrink the white caps of the Andes even further, and it could also increase evaporation from the bofedales, as well.

Secondly, although most models indicate wetter conditions in the region by 2100 AD, most of Peru's mountains are currently undergoing a drying trend that might last for several more decades, and that drying process can affect local wetlands as much as it does glaciers. Drought among the bofedales could therefore threaten alpine pastoralists and their alpaca herds as well as lowland farms and settlements, and an unpleasant combination of warming and drying might drive some people out of the highlands altogether.

In any system shift that is as all-encompassing as global climate change, there will be winners as well as losers, and the future of the low latitudes need not be totally grim. As Hadley circulation speeds up, for example, the ITCZ should drop more rain on most of the tropics. In some situations, heavy rains can be troublesome because they wash roads and bridges away, erode topsoil, and leave puddles for disease-bearing insects to breed in. But rain is also a valuable commodity in seasonally dry regions, particularly in nations that rely heavily upon agriculture and hydro power for subsistence and commerce.

Most Pakistani farmers or Indonesian rice-paddy harvesters are less likely to bemoan the arrival of heavy summer monsoons than to worry that the rains may weaken or fail on occasion. According to most global circulation models, however, the vast monsoonal regions of southern Asia will become wetter as the tropical atmosphere and oceans warm. More copious rains could also help to keep a sizable fraction of Amazonian rain forests green, and increased precipitation in the Andes later in this century might help to keep some ice and snow on the peaks there. Wetter conditions in the East African headwaters of the Nile could open new areas along the river's green flanks in lowland Sudan and Egypt for irrigation and settlement. Subsistence farming keeps many Kenyans and Tanzanians fed today, and just as the last bits of arable land are coming under the hoe we learn that more abundant and reliable rains—and fewer killing frosts at higher elevations—may be coming. Many who stand to benefit thus from such changes are among the world's least wealthy, and the possibility that warming might in some cases improve their lives should be considered seriously in any tally of the costs and benefits of Anthropocene climate change.

Unfortunately, such conclusions are difficult to make with perfect certainty. One can reasonably assume that warming should stimulate Hadley circulation and use that assumption as the basis for drawing rough maps of future wet and dry zones even without the aid of computer models. But again, the story may be more complex than it seems at first. Take cyclones, for example. In theory, warming should increase their frequency and strength, but despite well-documented heating trends in the lower latitudes there is blessedly little evidence (yet) for thermally enhanced cyclone activity in most of the topics. The same cannot be said for Atlantic hurricanes, though; several studies do indicate a significant increase in hurricane activity since the 1970s and 1980s that may be due, at least in part, to oceanic warming.

The uncertainty runs deeper still, because inherent variability in the Hadley cells and ITCZ are not the only players in the field of tropical rainfall dynamics. For reasons unknown even to solar physicists, the sun releases slightly more energy than usual every eleven years or so, and a growing body of evidence suggests that those mild fluctuations can sometimes influence weather on Earth. In 2007, my colleagues and I published a paper in the *Journal of Geophysical Research* showing that those small oscillations in the strength of the sun's output appear to trigger unusually heavy precipitation in East Africa, and that such rhythmic, rainfall-driven pulses occurred in the levels of Rift Valley lakes throughout the twentieth century. Similar decadal rainfall pulses have also been found in parts of southern Africa, only in reverse; there, the mild solar energy peaks are linked to droughts.

The effects of the equatorial rainfall pulses are complex, and not all of them are desirable. More rain can mean more water for crops and reservoirs, but it also spreads disease by creating puddles for mosquitoes to breed in. With clockwork regularity, episodes of unusually heavy rain have triggered major outbreaks of Rift Valley fever in Kenya since detailed record keeping began in the 1950s. Unfortunately, exactly how this apparent sun-weather linkage works is still unclear—most likely it is related to surface temperatures in the Indian Ocean—and to my knowledge no climate models have reproduced it yet.

But there is yet another source of tropical rainfall variation to consider, one that links the weather of Africa to that of Peru and other locations around the world. Its influence is more widely felt than the effects of the solar cycle, but its pacing is far less predictable and we know less about it than we should if we are to plan effectively for a warmer tropical future. That wild card is El Niño.

Shortly before global warming invaded the world's newsstands a decade or so ago, El Niño was the dominant climatic star on stage. If you're old enough, you might remember when it first drew extensive media coverage outside of South America, back in 1983 when one of the most powerful El Niño events on record threw weather systems around the globe into disarray. Perhaps it was the intensity of that disturbance, which drenched the southern United States and desiccated islands of the western Pacific, that first attracted world attention to what was previously treated like a quaintly localized phenomenon. Perhaps it was also the writings of prominent scientists who happened to be monitoring Peru's coastal upwelling system when the sudden slackening of winds shut it down. Whatever the reason, many of us are now more or less familiar with the term itself, which refers to the climatic disturbances whose onsets approximately coincide with December, the traditional birth date of the Christ child, usually every three to seven years or so.

El Niño begins with a slackening of easterly winds, which slows the upwelling of deep, cold waters along the coasts of Peru and Ecuador. As the sea warms, air moistens with marine water vapor and heats up enough to rise and condense into rain clouds. The effects of that regional switch from desert to deluge cascade through sensitive sites around the world, generally making the weather wetter in places that are normally somewhat dry (Kenya, Texas) and drier in the wet places (Queensland, Zimbabwe). In 1997–1998, when another severe Niño drought struck Indonesia, dehydrated peatlands caught fire and smoldered for months, choking everything from Jogjakarta to Singapore with lung-searing smoke. As much as a third of Indonesia's wild forest orangutans are thought to have died as a result of the flames and fumes, either directly or from forced migrations into dangerously unfamiliar or more heavily settled areas.

Such changes in weather patterns could be powerful enough to complicate expected rainfall trends in much of the tropics. But despite El Niño's huge influence on climates worldwide, we're still not sure what has been causing it since the modern version of it began, roughly 5,000 to 7,000 years ago, much less what it may do during the rest of the Anthropocene. Not surprisingly, computer projections about the coming responses of El Niño to rising temperatures vary a great deal. In a paper published in 2005, British climatologist Matthew Collins and a team representing sixteen international research groups reported that models in which the El Niño system changed the most also produced the least reliable simulations, and they concluded that no major disruptions of that system are likely to occur as a result of future warming.

Geologic history also paints a rather blurred picture in this case. Sediment cores from the Galápagos Islands, the Ecuadorian mountains, and marine deposits off the Peruvian coast contain abundant evidence of past El Niño floods, but the records that we currently have on hand still leave much unexplained. Most of them do suggest that a long-term suppression of El Niño activity occurred during northern summer warmings 7,000 to 10,000 or so years ago, but that change was due to factors other than human-produced greenhouse gases, and some of them offer conflicting versions of rainfall conditions during warm and cool phases of the last millennium. Furthermore, the prolonged global warmth of the Eocene epoch 34 to 55 million years ago caused little or no change in El Niño activity, at least none that has left convincing signs in published geohistorical records. This unfortunate situation leaves us largely ignorant of what El Niño will do to tropical rains in the future as we move along the rising and falling curve of fossil carbon emissions.

Although we can't be sure exactly how Hadley cells, El Niño, and related weather systems may respond as global average temperature rises, we can still be fairly certain that changes are imminent, if not already under way, and that some of them could be severe. With this in mind, we may pity official decision makers in tropical nations who hear the warning cry to "Look out!" as they struggle to see what lies ahead. A recent editorial in *Nature* described that situation at a conference in Johannesburg in 2005. A gathering of more than fifty foreign researchers warned government officials and planners that they need to "take urgent action" in the face of global warming, and said that "the key is to convert these concerns into action." But none of those experts explained what that action should be. How can you respond appropriately when you don't know what you're supposed to respond *to*?

It is often said that inaction in the face of impending climate change is not an option, but choosing the wrong response can be dangerous, too. In 1997, for example, meteorologists used sophisticated computer models to warn that a severe El Niño–induced drought was about to strike southern Africa. As a result many local farmers held off on planting crops that seemed to be doomed to failure, but the rains fell almost as abundantly as usual despite the forecast. When the nation's food supplies shrank in response to the farm cutbacks, people's trust in the meteorologists shriveled, too. That, in turn, caused a second wave of damage when resentful citizens later ignored subsequent warnings of impending floods, which actually did appear as predicted and killed hundreds of people in the region.

Rather than try to predict specific climatic changes with great precision—which many scientists believe is impossible even with the most sophisticated models—an increasing number of experts favor preparatory strategies that enhance general adaptability in anticipation of a wide range of possible futures. That approach typically involves building resilience by addressing vulnerabilities in aspects of life that are more immediately troublesome than global warming, such as poverty, disease, war, and limited access to education and technology. If you're well off financially and socially, then it doesn't matter so much what the weather is like; the example of oil-rich Dubai, on the Persian Gulf, shows that one can theoretically enjoy a comfortable lifestyle even in a barren sun-baked landscape where summer air temperatures exceed 106°F (41°C). It is mainly low incomes, societal instability, and inadequate infrastructure that make climatic change of any kind such a threat to so many residents of the tropics. Though some activists claim that adaptation is a cop-out, that all efforts should be concentrated on reducing carbon emissions as the root of the problem, I find it difficult to justify that attitude in the case of tropical nations. These include some of the world's least resilient cultures and economies, and they have made the smallest contributions to planetary heating in the first place. It is particularly important for them to maintain stable, well-balanced stances while walking into a poorly understood climatic future so they can react quickly and effectively to unforeseen obstacles in their path.

In that spirit, the following advice may be helpful: focus on the long term. Don't fall for illusions of stability or miragelike trends that can appear on short-term horizons. Remember those misguided blind men for whom the cursory feel of an elephant's leg, trunk, or flank made the creature before them seem to be a tree, a snake, or a wall; incomplete information can fool us into making dangerously inaccurate conclusions about the world we live in.

In much of Peru, for example, the accelerated melting of mountain glaciers is now sending extra meltwater downhill to thirsty farms, towns, and power dams, but the surplus will likely reverse into a deficit as the ice shrinks further, probably within the next few decades. Adapting to this temporary bonanza by boosting demand could therefore be disastrous in the long run. Better to concentrate now on conserving high wetlands and on building more reservoirs and canals, keeping a wary eye on a drier future (as well as on earthquake-resistant construction methods).

In Africa, the desiccation of Ferguson's Gulf need not have taken anybody by surprise if they had maintained a historical view of the place that was farsighted enough to recognize the full range of lake level variability. But a similar mistake is now being made in reports of a recent drying trend farther south in the Lake Victoria watershed. The level of that lake has been falling since the 1960s, which threatens to strand existing fishing boat launch sites farther and farther from the water's edge and weaken the generative capacity of hydropower turbines on the Nile outlet. But this is probably not really a sign that global warming is sucking the lake basin dry. Charts of the full twentieth-century lake level record suggest that this short-term trend is part of a decades-long recovery from an as yet unexplained, unusually wet period that soaked tropical Africa between 1961 and 1964. In that case, it probably has little predictive value other than to serve as a reminder that tropical rainfall regimes can be extremely variable; if anything, the equatorial Victoria basin should eventually become wetter as it warms through the twenty-first century and beyond, not drier.

A truly long-term view of climate also warns of another aspect of Anthropocene change that favors thoughtful flexibility rather than rigid, overly specific preparatory strategies. Many of the trends that are appearing around us now will eventually reverse. After thermal maximum passes and climate whiplash runs its course, inexorable global cooling should progressively slow the Hadley circulation down, thereby weakening monsoons in much of the tropics and lightening up on some of the aridity elsewhere. The crest of Peru's snowy Cordillera Blanca may spend some time looking more like the "Cordillera Dry and Brown," but then the increasingly frequent frostings that drop from alpine skies will take longer and longer to fade away, and they will hold the promise of heavier, more reliable snowmelts to come in the deep future.

On the far, cooling side of the coming thermal divide there will be new environmental winners as well as losers, just as there are now. To paraphrase Bob Dylan, "the first ones now may later be last." Best to keep our eyes open and focused far ahead because some of the chances available to us now for effective preparation and adaptation may not come again. Yes, tropical climates are changing, too; of that, and only that, we can be sure.

11

Bringing It Home

Not all change is bad. Where a reasonably healthy, reasonably diverse ecosystem is providing at least some kind of service, we might be better off to embrace our altered Earth. . . . We may even learn to find some charm there.

—*Editorial,* Nature, *July 23, 2009*

Most of this book has been dedicated to the understanding of grand sweeps of time and planet-spanning environmental changes. But the events that it describes will be experienced by our descendants on much smaller, personal scales. Global averages of temperature and precipitation are constructed from individual data points, as a house is built from various boards and beams. Each component has a more or less unique shape and position, and the dimensions of any given handful of them can differ widely from the overall average. The computer models that illustrate global trends are too farsighted to focus easily on the little places where you and I and future generations must deal with climate up close. And in order to envision what greenhouse warming means for specific locales we need more place-based information to guide us.

In this chapter, I'll turn to the temperate zone for examples of what the Anthropocene may have in store for some of those little places that fall between the climatic extremes of the poles and the equator. At the highest latitudes, many of the most important ecological changes will involve the melting of ice, and in the tropics, rainfall variability will be the dominant theme. But in the middle latitudes, those that encompass the United States, central Eurasia, and the southerly portions of Canada, South America, Australia, New Zealand, and Africa, the situation is rather more complex. In those regions, ice and snow are confined to higher elevations and to certain times of year, and precipitation or drought are rarely limited to any particular season.

As a result of that geographical diversity, it can be more difficult to anticipate what Anthropocene warming means for a temperate-zone site than for the North Pole or humid equatorial Africa, other than a near-universal rise in mean annual temperatures. To do so with a reasonable amount of certainty requires intimate knowledge of the setting in question and an awareness of how changes in planetary-scale weather systems influence the small-scale locales that we actually live in. Such place-based familiarity is essential to understanding what global climate change means to one's home turf, and in that spirit I'll choose two such places here for illustrative purposes, the Cape Province of South Africa to represent the southern midlatitudes and upstate New York to represent the northern ones. I've selected these locales because I know them fairly well and because they represent two very different climatic regimes, with the African site being more vulnerable to coming changes in precipitation and the American one slated to be more directly affected by warming. But I'll focus most closely on the latter region for one simple reason: I live there.

It may seem odd to some of us that the southern tip of Africa lies in the temperate zone. In the austral autumn, brown leaves fall from the sycamore trees and crunch underfoot on the sidewalks of Cape Town. In winter, it often snows on the high rocky escarpments that loom over the vineyards of Paarl and Stellenbosch. For much of the year, many South Africans wear sweaters and pull knit caps down tight over their ears while most of the rest of the continent swelters in tropical heat. But when summer comes, the seasonal warmth is glorious and the long sandy beaches draw crowds of swimmers and sunbathers. In some ways the weather resembles that of the semiarid Mediterranean or Californian coasts, but one is quickly reminded of South Africa's uniqueness when one looks closely at the native vegetation out in the open countryside.

Superficially, the Cape's unusual fynbos may resemble the succulent, aromatic plant communities of dry habitats elsewhere, but there's nothing exactly like it on Earth. The ancient plant assemblages that grow here evolved in isolation from other temperate regions and are incredibly biodiverse. A few minutes' stroll atop Table Mountain, a short cable car ride up the sheer gray cliffs from Cape Town, introduces you to a rich array of otherworldly plants. Leafless epiphytes that resemble bright orange string hang in thick tangles from *Protea* bushes that produce woody cones and glorious colorful blossoms. A so-called three-day-blister bush resembles celery but raises vicious welts long after it's touched; fortunately for hikers, it's readily avoided if you keep a wary eye out for it. Tiny, insect-eating sundew plants no larger than a fingernail cling to dense cushions of green moss in the wetter areas. And the list goes on, adding up to thousands of species that are found only in this small patch of temperate habitat on the dangling tip of Africa roughly halfway between the equator and the South Pole.

But it's not just the physical isolation by tropical heat and deep-ocean waters that makes the fynbos into the unique botanical collection that it is. Rainfall plays a role, too. Any plant species that tries to join the

Fynbos plants at the Cape of Good Hope, South Africa. *Kary Johnson*

thriving vegetational community on Table Mountain has to put up with winter chills and nutrient-poor sandy soils, but also with prolonged drought. It rains so infrequently there that water conservation measures are necessary for survival. These may include waxy coatings on foliage, liquid storage in juicy stems and fleshy leaves, or seasonal dormancy. But even these adaptations don't make for easy living, and extra water rations from cool, damp fogs that drape the crests of the hills also help the hardy plants to make it from one rainy period to the next.

At tropical latitudes, rain is normally most abundant when summer heating drags the seasonal rain belt over the region, but in the South African Cape region there is another climatic mechanism at work. Westerly winds dominate the weather there, just as a matching stream of mid-latitude westerlies pushes storm systems from west to east across the United States, Canada, and Eurasia. In most places in the northern and southern temperate zones, westerly winds can carry clouds overhead at any time of year, but in southernmost Africa it mainly rains in winter.

By an accident of geography, the Cape region lies just barely within reach of ocean-fed rainstorms that ride the winds across the South Atlantic, and they come ashore mainly when winter cooling over Antarctica drives the encircling wind tracks farther north than usual. When the westerlies shrink back poleward during the warmer months of the year, most storms miss that marginal landfall and simply continue on across the southern rim of the Indian Ocean. As a result, the Cape experiences one major rainy season (winter) and one long, mostly dry season over the course of a year.

What does Anthropocene warming hold in store for this place? Sea-level rise will gradually push the famous beaches farther inland and allow salty water to creep farther into coastal estuaries and settlements. Warmer temperatures will increase evaporation rates during the dry seasons and therefore add to the inherent aridity, and less of the winter precipitation at higher elevations will fall as snow. But these are not the only changes to come. Warming will probably hold the winter storms offshore for longer and longer periods, shrinking the duration and intensity of the rainy seasons more and more. Such a linkage between warming and drying is revealed in sediment core records of the last millennium that my associates and I have recently collected from South African lake deposits, and in a much warmer future, the winds may shift so far south that winter rains rarely strike the Cape at all. While most of Africa becomes wetter as global temperatures rise to thermal maximum, southernmost Africa is likely to take a different path and become drier than ever.

Mike Meadows, a paleoecologist at the University of Cape Town, worries that if winters become shorter and drier, then South African water supplies could dwindle and cause problems for farms and towns, the wine industry, and the unique fynbos. "These plants can't just lock arms and move all together to wetter locations," he explained during my latest visit to his laboratory. "They're tough and resilient, so some species may pull through just fine. But more drought-sensitive species might be lost."

It's not just thirst that threatens the plants, most of which are already used to very dry conditions. It's fire. "Less rain in the winter would lead to even drier soils and the woody vegetation would be more flammable in the windy summer when fires usually occur," Mike continued. "This may lead to hotter fires that can cook the soils so much that they actually repel water when rain finally does fall." Seeds would struggle to sprout, and the runoff would wash valuable topsoil away.

Look around the rest of world's temperate regions and you'll find similarly unique, place-based stories unfolding in different locations. In southern Australia, as in southern Africa, people are worried about future drying and fire hazards. In the Alps, skiers and climbers are watching their beloved snow and ice fields shrink, and woody shrubs are moving uphill into alpine meadows. In the high, frosted Himalayas, a vigorous debate is under way among scientists; some fear that glacier retreat could bring water shortages to millions who live downstream, but others find little sign of such retreat and argue instead that the primary source of lowland river water is monsoon rainfall, not meltwater. In much of China, rain is expected to become more abundant overall, but the intensities of floods and droughts appear to be increasing. And nations that rim the Mediterranean Sea are concerned about thick, bottom-smothering globs of marine slime that are building up in the increasingly stratified waters, apparently more in response to rising temperatures than to other human influences.

But there are surprises in the mix, too, a reminder that globally averaged patterns don't capture the diversity of local-scale situations on the ground. A paper published recently by climatologist Alexander Stine and colleagues in *Nature* reported significant winter cooling in parts of Quebec and the southeastern United States between 1954 and 2007. This is seemingly at odds with the global situation, but it doesn't negate it, either. A worldwide average is a composite of many locations, some of which may differ quite a bit from the mean. With Earth as a whole so obviously warming up, finding subsections of North America on a cooling jag can simply mean that other parts of the world must also be warming faster than the average (the western Antarctic peninsula, for example).

In my own home region, the Adirondack Mountains of upstate New York, our main concerns differ from those of the dry South African Cape. Water is abundant here, and with the majority of computer models predicting slightly wetter conditions for us in the future, we tend to focus more on temperature-related changes and try to anticipate their effects on the tourism and winter sports industries that help to sustain our local economy.

The Adirondack State Park centers on a Vermont-sized dome of ancient anorthosite rock capped by rugged peaks and sprinkled with tree-lined lakes. New York's tallest mountain, our mile-high Marcy, towers over the northernmost headwaters of the Hudson River. Manhattan is a six-hour drive due south, and Montreal is less than three hours north. About half of the park's 6 million acres are privately owned, and our patchwork landscape is a unique, invigorating blend of wilderness and humanity. I live and work in the northern sector at Paul Smith's College, a small, rural school on the shore of Lower Saint Regis Lake, named after a nineteenth-century wilderness hotelier and entrepreneur. It's an idyllic place to call home, and I enjoy it when people ask me what town I live in. I reply, "There *is* no town where I live."

Although I study climate change, most of my research deals with the distant past in general and the tropics in particular. Therefore, I didn't spend much time studying Adirondack climates at first. In that sense, I was much like other concerned citizens who get most of their breaking news from the media. All I knew about local climate was that the weather around here seems to be unpredictable, that it's bloody cold in winter, and that our fiercest summer heat waves rarely hit triple digits.

As global warming began to seep into public awareness during the last couple of decades, it was only a matter of time before someone published a compelling description of life in a hotter world that could be tailored to this popular neck of the woods. Bill McKibben, who lived in the central Adirondacks at the time, was the first writer to do so in a big way in 1989 with his groundbreaking book *The End of Nature*. His work gave an early preview of a possible future here: a once-diverse forest of mixed northern hardwoods and conifers devastated by warming, a place that rains instead of snows in January, a former wilderness disfigured by the stain of carbon pollution.

At first, I was cautiously skeptical of such claims, as many of my paleoclimate colleagues were at the time. For us, climate change was about ice sheets bulldozing entire ecosystems off the map and natural hothouses keeping dinosaurs comfortable in the distant past. A couple of degrees warmer by the end of this century? Ha! That's nothing. And how do we know that it's really happening *here*? Global trends need not represent our local patterns accurately; show me the data for *this* location, and then I'll believe you. Regardless of whether such a stance proves to be correct in the end, it's the basis of healthy questioning; that is to say, the time-tested way of science.

As luck would have it, I met and befriended Bill McKibben before I ever saw those convincing data. As a well-known park resident he was invited to join my college's board of trustees during the 1990s, and he not only accepted the offer but took the post seriously enough to become personally familiar with our little school-in-the-woods. I enjoyed getting to know him, but I still wasn't convinced by some of his claims of impending trouble. As a scientist, I wanted to see the numbers behind them but knew of no published Adirondack-specific weather data that could confirm or reject the existence of a major warming trend here.

That situation soon began to change. Teams of scientists from various state and federal institutions were busy ratcheting the global picture down to the scale of national subregions by compiling weather records, evidence of changes in water bodies and woodlands, and more finely focused computer simulations to track regional climate trends and project them out to 2100 AD. If you live in the United States, then you may well have such a regional assessment available for your home region that can be found through a local university or with a simple online search.

In 2001, one such group that was based at the University of New Hampshire published their findings in a report titled "New England Regional Assessment" (NERA). The group spread their findings widely in print and in person, touring to speak throughout the northeastern states. But despite their efforts most Adirondackers still encountered those findings mainly through interpreters such as Bill, who described some of NERA's results in *Adirondack Life* magazine amid other gripping descriptions of our future. In that article we learned that warming is already under way here, that it will soon kill our sugar maples, replacing their brilliant autumn oranges and reds with the somber browns and dull greens of oaks and hickories that are now more typical of the Blue Ridge or Smokies, and that our winter sports industry will soon collapse as snowfall turns to rain.

For those of us who love the harsh beauty of the North Country, imagining a winter without snow is like imagining cake without frosting. Winter here means critical tourist dollars, a shot at hosting the Winter Olympics again, white Christmases, skiing and snowshoeing and skating, an icy closing of the annual circle of the seasons, and a cherished prelude to springtime. Simply put, Adirondack winters are *supposed* to be snowy.

In the hands of such a master wordsmith, this kind of imagery hits us hard in the gut. But Bill didn't come up with the information himself; he merely handed it over to us in beautifully crafted prose. If you've been getting your own previews of impending change in your home region by this sort of indirect route, then you may also want to dig deeper into original sources to judge their accuracy.

Bill gave me a chance to do that by asking my opinion about local climate trends as he prepared his *Adirondack Life* article. My response to his invitation was a welcome personal project that gave me an excuse to see how well NERA's work stacked up against what I could find on my own. They hadn't singled the park out for specific analysis and had instead lumped it in with other parts of New York state, but I suspected that this complex, mountainous area might not be following other places in perfect lockstep.

I began by digging through my collection of geohistorical literature and soon found evidence to support the possibility of oaks and hickories dominating our forests in a hotter climate. In 1993, Syracuse University geologist Ernest Muller and colleagues published a description of ancient lake sediment from the warm Eemian interglacial that was exposed by a large mining pit at Tahawus, in the central Adirondacks near Mount Marcy. It was sandwiched between two layers of glacially deposited sand and gravel, and pollen grains extracted from it showed that oak, hickory, chestnut, black gum, and beech dominated the woods back then. All but the last kind of tree are common in the southern Appalachians but rare to absent in the Adirondacks today.

If our greenhouse gases bring Eemian-scale temperatures back to these mountains, which appears likely even in a moderate 1,000-Gton emissions scenario, then history supports predictions of southern-style woodlands returning as well. That still doesn't necessarily mean, however, that it's going to happen during this century, or that such a change would be all bad from the point of view of future Adirondackers. Oaks and hickories have already lived here in abundance, so in the strictest historical sense they're as native to these mountains as today's trees. Oaks make acorns, which could help to keep bears, deer, squirrels, and other wildlife fed despite the current decline of beechnuts caused by beech bark disease. And I've traveled to the mountains of western North Carolina in order to enjoy the autumn colors with friends. It wasn't as spectacular as what a more densely maple-studded landscape can offer, being rich in oranges, golds, bronzes, and muted reds but without so many of the brilliant scarlets and tangerine bursts that enliven more northerly scenes. Nevertheless, those colors still draw swarms of appreciative leaf-peepers to the southern Appalachians every year.

My next task was to find out if the Adirondacks have really been warming as fast as NERA had claimed in its summary of northeastern climate trends. I enlisted the aid of a colleague, Mike Martin of Cedar Eden Environmental Consulting, who gathered the daily records of eight Adirondack weather stations from a database maintained by the National Climate Data Center. Several of the records showed moderate overall warming during the last fifty years and some showed slight cooling, but for all of them the most pronounced feature was their variability from year to year, month to month, and even day to day.

No great surprise there. NERA's own maps showed much of Maine

cooling slightly while the rest of the northeast warmed, and the Adirondacks have some of the least stable weather conditions of any similar-
sized patch of the lower forty-eight states. Local residents know this
variability well, though they didn't make a big point of telling that to outsiders as they coaxed a second Winter Olympics back to Lake Placid in
1980. Some readers may remember the January thaws that almost can-
celed the event that year. Having just helped to glaze the serpentine bob-
sled track at Mount van Hoevenberg with ice, I watched in frustration
as it all melted away shortly before competition began. Fortunately, a last-
minute freeze and timely snowfall saved the games in the end.

An important discovery soon emerged from our number-crunching
exercise, one that you'll probably make yourself if you examine weather
records from your own home region. Mike and I could find any kind of
trend we wanted to amid the short-term ups and downs of interannual
temperature fluctuations. It all depended on the time scales we chose to
look at.

For instance, let's say that we wanted to support a frightening end-
of-winter story. We could then choose a subset of time that began with a
cool period and ended with a warm one. The 1960s were relatively cool, so
we could simply focus on the last four decades of the twentieth century in
order to highlight an ominous trend rising from low to high temperatures.
On the other hand, if we wanted to seem like pesky contrarians, we could
instead show that a longer, fifty-year interval displayed a slight cooling
trend overall. That's because the early 1950s were unusually warm here, at
some sites even warmer than the last decade was. A similar warm-cool-
warm pattern is found in many other weather records from the northern
temperate zone, and it helps to explain why some locations show average
cooling over the last five decades. It's like placing a long plank atop two
widely spaced boulders; if one stone is larger than the other one, the board
(trend line) will dip downward (cooling) toward the smaller stone. Choosing the hot 1950s as a starting point therefore obscures a more recent pattern of renewed warming that began in the 1970s.

And another discovery soon followed. The midcentury Adirondack
records differed quite a bit from the global pattern. Global average temperatures jumped during the 1940s, but the Adirondacks cooled slightly
then, and the pattern reversed a decade later to produce that troublesome

local warm spell of the 1950s. All the more reason to determine what the climate of your home region is really doing rather than assuming that it always operates in lockstep with the global average.

I passed my results along to Bill, who reported some of them in his *Adirondack Life* article. They didn't resonate well with NERA's findings, though, and I smiled to see myself introduced to readers as one of the "few scientists left who aren't convinced the climate is going to change dramatically." But a much harsher response came after Mike and I published our results, which showed little or no fifty-year warming trend in local weather records. The article, which appeared in the *Adirondack Journal of Environmental Studies*, got us lambasted in the local paper by the head of an environmental group and a member of NERA itself, who apparently suspected us of being climate naysayers. The blowback surprised me, but it also goaded me to examine the Adirondack situation even more closely. As I dug deeper, I began to understand why it's so difficult to find reliable descriptions of local-scale climate change in the mainstream media.

By tapping into the professional grapevine, I soon found that scientists from the University of New Hampshire, Middlebury College, and the U.S. Geological Survey were faulting the NERA report for its reliance on data that yield "spurious temperature trends." Applying more carefully vetted weather records to the same region, those scientist-critics found more widespread warming than the earlier report did, and some of the differences were quite striking. For example, NERA's finding that most of Maine cooled off during the twentieth century conflicted with the newer data, which showed that more of it had, in fact, warmed along with the other New England states.

Although the new studies knocked bigger holes in NERA than Mike and I had, they didn't leave us gloating. The dataset under attack was the same one that we had obtained from the National Climate Data Center, which meant that some of our own results were wrong, too.

The mistake was innocent enough, though regrettable. Rather than try to sift local climate trends out of a broader study conducted by others, we had gone directly to the raw weather measurements. But what we, the NERA team, and quite a few other investigators didn't fully appreciate at the time was that this approach produces a host of more or less random errors in the magnitudes, and even the directions, of climatic trends.

Thus chastened and newly enlightened, I was led to the straight and narrow path of good clean data by Jerome Thaler, a climatologist from southern New York whose book *Adirondack Weather* caught my attention from the shelf of a local bookstore.

"The first thing you have to realize," he explained over the phone, "is that these data are collected by real people, most of whom are unpaid volunteers." In other words, the individual habits of people running a station can strongly affect the data that they collect.

For instance, Jane runs the local weather station for twenty years. She awakens early each morning to record the temperature before heading off to work—unless she's on vacation or the kids are sick. That leaves a gap in the daily readings and distorts the monthly average temperature calculations.

When Jane retires, John offers to take over her duties. But John doesn't like to get up early, so he takes his readings later in the morning when the sun starts to warm things up for the day. Automatically, and incorrectly, the daily temperature averages become warmer.

And then there are equipment upgrades, power outages, changes in the numbers of readings per day, changes in station location, and changes in local vegetation, all of which can affect temperature data. If those factors aren't recorded carefully along with the weather measurements, then there's no way to correct their influences on the true temperatures.

"When climatologists say that they've cleaned or adjusted a weather record, it's not cheating, as some might claim," Thaler continued. "You simply have to do it in order to correct the errors as accurately as you can." The current best source of weather data, according to Thaler and every other expert on my contact list is the U.S. Historical Climatology Network (USHCN), the same source that was used to challenge NERA's findings.

USHCN staff select only those stations that fully document such error sources. They correct for informational gaps as well as changes in physical settings and methodologies, and they explain their methods in deadly detail on a website that offers regularly updated data in chart or table form.

And what do the new and improved USHCN records tell us about the Adirondacks? Several local weather stations have still cooled slightly since the 1950s, but the trend at Wanakena, in the northwest corner of the park, no longer dropped as it did with our raw dataset. Averaging all of the Adirondack station curves into one bouncy line showed the ups and downs balancing out into a slight overall rise of mean annual temperature during the twentieth century, but an upswing since the early 1970s warmed the whole region more steeply, by about a degree and a half, thereby more closely echoing the global pattern. The change was most significant, in a statistical sense, during June and September. December temperatures also rose, but they did so too erratically for the trend to be distinguished from random variability, and the other months of the year showed no major trend one way or the other.

Okay, now I believe it. We're warming up here in these mountains, too. But how fast? If you look only at the last fifty years, then you see a much weaker average warming than what you find if you examine the last thirty years. So how should we decide which time period to use?

Most experts define a region's present climate with a sliding time window that averages the last thirty years of weather conditions. For example, a scientist speaking in 1990 would have described the climate of the Adirondacks by summarizing the previous thirty years of weather data spanning 1961 to 1990. Someone doing the same thing a decade later would have examined data from 1971 to 2000, and so on. Therefore, the last three decades are a reasonable time frame to focus on for a study of local climate. In addition, reputable studies show that the post-1970 global temperature rise is more clearly attributable to greenhouse gas buildups than to any other factor, such as changes in the sun's output or clearing skies. By those criteria, we can justifiably treat a recent thirty-year average warming trend as a sign that global change is probably at work here like a slowly rising tide beneath the erratic chop of short-term weather fluctuations.

In 2006, a new team of investigators from several northeastern institutions published a report entitled "Northeast Climate Impacts Assessment" (NECIA). To my relief the regional weather histories in the report more or less resembled my own updated Adirondack reconstructions. So, too, did a subsequent analysis of upstate New York weather records that was conducted as a master's thesis project by Kathie Dello at the State University of New York, Albany. NECIA's work still made no Adirondack-specific predictions, but knowing that both local and large-scale climates really have begun to change in similar fashion during the last three decades made their conclusions and my own seem more reliable.

It took a lot of work to convince me that the Adirondacks are warming. Not because it seemed particularly unlikely, but simply because there was little solid evidence available to support that claim until recently. Furthermore, it can be difficult to link local changes conclusively to the global greenhouse effect because the large-scale trends that it produces are slow and gentle compared to the erratic flip-flops of short-term, small-scale weather, especially up here in these mountains. It's a shame that those who deny the existence of global warming out of stubbornness or willful ignorance so often bear the epithet "climate skeptic" (rather than something like "naysayer" or "contrarian"), because the job of a good scientist is to maintain reasonable skepticism in the face of a story that's long on sex appeal but short on facts. Until someone downscales global climate change information to fit your particular piece of the planet with demonstrable accuracy, you can't really be sure whether it moves in the climatic mainstream or not.

Here in the Adirondacks, though, the long-term warming trend that is now firmly documented by local weather records is already producing some clearly observable changes in the realm of winter ice. According to the NECIA report, ice-out is now coming earlier to lakes throughout the northeastern states than it did a hundred years ago, though that's not necessarily temperature at work. Snow and wind also control the fate of lake and river ice. Melting happens later in spring if the ice is well insulated by a blanket of snow, and that blanket thins or thickens depending on how much snow falls, how much melts, and how much blows away. And what usually makes lake ice break up at any given moment is wind, not heat alone. In calm weather, lake ice can sit for days, rotting into brittle, vertically packed needles, as it waits for the first breeze to smash it against the downwind shore.

A clearer indicator of warming is the date of freeze-up, because it is unaffected by the complicating effects of snow cover and midwinter ice thickness. Moreover, our local warming trend in autumn far exceeds that in spring, and this seasonal imbalance is reflected in the behavior of our lake ice. Thaler's book presents a century-long record of ice cover from Mirror Lake, an attractive body of water in downtown Lake Placid, and local librarian Judith Shea helped me to update that record by contacting a boating club that runs an annual ice-out contest there. Those data show that Mirror Lake now freezes two weeks later than it did during the early 1900s; the more erratic ice-out record, on the other hand, yields only a weak trend toward slightly earlier dates. And at lower elevations, where temperatures are already warmer to begin with, the change is even more obvious. Lake Champlain, which occupies a long valley on the eastern border of the Adirondack Park, hasn't frozen over at all in recent winters. Records dating back to the early 1800s show that Lake Champlain skipped its winter freeze-up only three times during the nineteenth century but more than two dozen times since 1950. That's difficult to explain with anything but a warming trend.

I published these and similar results with several colleagues in a recent issue of the *Adirondack Journal of Environmental Studies,* and when the Associated Press later interviewed some of us about the regional re-treat of lake ice, I was amused by the public response that the article trig-gered. Almost all the comments that I found posted online were negative. One writer called it "a cherry-picking promotion for the discredited global warming movement." Another typical posting said, "Global warm-ing can be traced primarily to the fevered ravings emitted by ecofreaks."

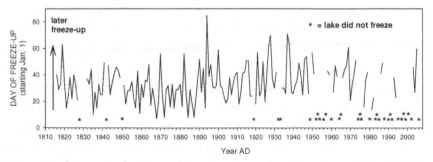

Freeze-up dates on Lake Champlain since the early 19th century. *National Weather Service Forecast Office, Burlington, VT*

Having previously been accused by the environmental community of not being alarmist enough, I took these opposing comments to be a good sign for a scientist. When you're bashed by both sides in an emotionally charged argument, I like to believe that it means you're standing in the middle where the truth probably lies. But the humor of the situation quickly faded when, just a few weeks later, three people lost their lives by falling through the ice on a local lake, and the U.S. Coast Guard warned snowmobilers and fisherfolk that the ice on Lake Champlain has become treacherously thin.

Another potentially useful source of information about the effects of climate change on specific locales such as the Adirondacks is less scientific in nature: the informal observations of amateur naturalists. A former student of mine, Brendan Wiltse, investigated the environmental history of Lake George for his senior thesis by comparing modern water quality conditions near his family's lakeshore home to those of the past as revealed in sediment cores and historical documents. One of his key sources of information was Thomas Jefferson, who visited the area in 1791. Jefferson kept written accounts of his natural history observations during that visit, which show what the beautiful lake and surrounding forests were like two centuries ago. "Its water is limpid as crystal," he wrote, "and the mountainsides are covered with rich groves of fir, pine, aspen, and birch down to the water's edge." Murkier conditions today in some parts of the lake therefore warn of a troubling decline in water quality due to recent human activity. Unfortunately, Jefferson's visit didn't last long enough for him to record any useful weather observations that we could also compare to present conditions.

Records of ecological change in more modern times can be helpful, too, thanks to both professional and amateur naturalists. Since 1991, I've been recording when the first robins arrive on the Paul Smith's College campus, when the native solitary bees emerge from their burrows on the sunny, south-facing slope of Essex Hill, when the fiery red buds burst on the maples that grow next to Cantwell Hall, and when the wood frogs and the black-and-yellow-spotted salamanders migrate to their ephemeral melt-water breeding pools on Keese Mills Road. This has become something of a ritual for me over the years, a way to feel more closely connected to the other species living around me and to better sense the rhythm of the seasons. And, of course, it has also allowed me to watch for local signs of change, which is why I'm careful to make my observations in precisely the same locations every year.

Most of the creatures and plants on my watch list have made no statistically significant changes in when they show up, wake up, or open up since 1991, probably (as local weather data show) springtime temperatures haven't risen significantly here in recent decades despite the warming at other times of year; perhaps it's also because I haven't been watching long enough yet to detect subtle underlying trends amid the random variability. While there's little sign of change in my observational data, I wouldn't even have known about that lack if I hadn't kept watching for as long as I have.

But some North Country residents have been at this sort of thing for much longer than I have, as I discovered by putting the word out on our local radio station. For the last two decades I've cohosted *Natural Selections,* a weekly science program on North Country Public Radio with news director Martha Foley, and in May of 2007 we dedicated an hour-long call-in session to signs of climate change in the area. The response was both entertaining and informative.

A listener from Fort Covington called in to say that her elderly father has documented the return of redwing blackbirds to their favored cattail marshes every spring since 1969. According to his handwritten notes, the birds aren't showing up any earlier or later, on average; again, this is probably because spring temperatures here still show little or no sign of a greenhouse-driven trend.

Biologist Stacy McNulty from the Huntington Wildlife Forest passed along some records of *Trillium* and hobblebush flowering dates that began in the mid-1970s, along with ice-cover data from local lakes. No significant trends among the flowers or ice-out dates yet, but freeze-up is happening significantly later on the lakes.

Jeff Chiarenzelli, a geologist at Saint Lawrence University, sent in a compilation of discharge records from several Adirondack rivers that showed a significant increase of runoff during the twentieth century, especially since the early 1970s, that mirrors what I've found in local precipitation records.

And master gardener Danna Fast has been recording the blooming dates of wildflowers near her home since 1982. Her white water lilies now open about two weeks earlier in summer than they did during the early 1980s, perhaps because the water they float in is getting warmer; June temperatures have indeed risen significantly here during the last three decades.

As we look ahead to how future climate changes might affect our landscapes and living things, we have less aid from computer models than we do for climate itself. Living things are not as easy to simulate as air masses are, but some investigators get a lot of press for trying anyway.

Such is the case for the flameout of sugar maples described in Bill McKibben's *Adirondack Life* article. Bill got the story indirectly from other folks, who apparently got it from a study conducted by the USDA Forest Service. Their website displays colorful maps of expected changes in the optimal temperature ranges of various tree species, which, in turn, refer to a study that was published in 1998 in *Ecological Monographs*.

At first glance, the maps seem to warn of an imminent die-off. The tinted blobs representing sugar maples shrink away toward Canada, while oaks and hickories creep in to replace them. Maybe the Adirondacks really are slated to become botanical equivalents of the Blue Ridge and Smokies by 2100 AD.

But there's a prominent button on the home page and readers are urged to click on it. When I clicked recently, I read the following: "In order to avoid the misinterpretation of our atlas, we want everyone to read the following section before making sense of the maps. We would like to stress here that our model is not predicting migration of species x—but rather the movement of suitable habitat for that species."

In other words, these maps show preferred climatic conditions, not species migrations. Trees don't simply uproot and chase after warmer temperatures. They also need suitable rainfall and soil types, effective ways to spread their seeds over long distances, and sufficient elbow room among established trees that can live for hundreds of years. One review of the subject in *Frontiers in Ecology and the Environment* pointed out that this kind of model "ignores transitional periods in which species may persist outside their normal climatic range," and other studies have shown that human presence has altered the distributions of many species so much that it's now impossible even to know exactly what their true natural ranges are in the first place. At this point I knew that I needed to speak with someone who knows maples much better than I do.

I turned to Mike Farrell, a forest ecologist who directs Cornell University's maple research station near the ski jumps on the outskirts of Lake Placid. Mike has heard the stories about maples dying from global warming, but he doesn't buy most of them. "Nobody really knows what's going to happen here in the future," he began, "but as far as oaks and hickories replacing maples in fifty or a hundred years is concerned, I think it's unlikely. Sugar maples can actually grow better in warmer settings than the Adirondacks, and there's a perfectly viable sugaring industry down in West Virginia. The main problem is with regeneration of new seedlings." Acid rain, diseases, and deer-browsing already hurt the existing trees here and they also tend to keep new maples from replenishing the woodland ranks. According to Mike, those same problems might keep invasive oak and hickory seedlings in check, too.

These comments, though backed by years of experience, were so different from what I'd been hearing that I pressed further. Are our sugar maples really not seriously threatened by climate change?

"Some things certainly could change here as it warms," he explained. "We do see sap runs starting earlier in the season now, but that doesn't hurt the trees any. The main risk would be if the summers become a lot drier, because maples can't handle droughts very well."

That put the ball back in my court. I dug into my collection of Adirondack weather records to see what's been going on with our summer rains and what may lie ahead. The last century saw no apparent long-term shift in drought frequency, and the last thirty years were slightly wetter than the previous eighty years, but with no significant trend one way or the other in summer rainfall. So perhaps there is nothing particularly worrisome in terms of climate for maples here after all, at least in the near future.

NECIA used several computer models to predict that annual average temperatures will be as much as 12.5 degrees warmer in the Northeast by century's end under an extreme-emissions scenario, and that annual precipitation totals could be 10 to 15 percent higher, but there's no way to be sure how accurate such models are without waiting to see what really happens. Nonetheless, you can still run them through a few hoops to see how well they perform on a known playing field, and historical weather records are the best tools for that job. To meet the challenge, a model must run its predictive time clock backward and produce a curve that closely resembles the real records. NECIA tried that very thing—a process called "hindcasting"—and reported that their models did an acceptable job of reconstructing the last decades of overall warming, but they didn't make nearly as strong a showing in the precipitation trials. A multiyear dry spell during the 1960s, what people back then called "the great northeastern drought," cut a deep and memorable notch in regional rainfall records, but the computer reconstructions missed it entirely.

The failure to reconstruct the 1960s drought properly isn't very surprising, though. Rain and snow are more difficult to work with than temperature because they vary more in time and space. Greenhouse gases mix easily throughout the lower atmosphere and wrap the entire planet in a fairly homogeneous heat-trapping blanket, so it's relatively easy for a computer model to simulate their effects on temperature. And even without our carbon pollution in play, temperatures tend to be more smoothly distributed over a landscape than precipitation is. On a hot summer day, for example, everybody in a given region will swelter more or less equally, but not everyone will necessarily be soaked by a thunderstorm that happens to pass through the area. Random turbulence or a bump in the terrain can lift and cool a bubble of humidity so it drenches one place while another goes dry, and plumes of precipitation from local lake effects or various colliding air masses can strike one place but miss another. To deal with such complexity, you have to average more wet-dry data together than you do when documenting temperatures, and even then you still need to treat the results with care.

But the problems with predictive modeling can also run deeper than that. The brief ups and downs in simulations aren't always linked definitively to particular dates, which makes the basic trends in their long-term forecasts more reliable than the shorter year-to-year fluctuations. In addition, different models are built around different assumptions that oversimplify or exaggerate various aspects of the climate system in unique and potentially conflicting ways. And perhaps surprisingly, zooming in on smaller subregions of the planet doesn't necessarily simplify matters but instead can add even more complexity by forcing the models to deal with mountains, lakes, and other local weather-distorting features. Such downscaling also reduces the number of weather stations left within the study area to support the simulations with historical data, and it can magnify systemic errors that lurk within the global-scale models themselves. In a recent essay in *Nature* by correspondent Quirin Schiermeier, one expert grimly summed the situation up with "our current climate models are just not up to informed decision-making at the resolution of most countries."

One reasonable way to deal with this kind of uncertainty is to consider results from multiple sources rather than settling for just one. An excellent mechanism for doing this is Climate Wizard, a user-friendly online analysis engine that was developed by the Nature Conservancy, the University of Washington, and the University of Southern Mississippi. The website presents you with a map of the world that lets you focus on various regions of interest, and it allows you to harness any of sixteen well-known global climate models to predict temperatures and precipitation during the rest of this century. Climate Wizard's impressive stable of models can yield a wide range of conflicting results if you ask for detailed predictions about a particular season, a short stretch of years, or certain small locales. But the broad-brush generalizations are remarkably consistent, as if the models prefer to "think big" rather than being forced to squint closely at an overly restrictive time or place. For example, all of them anticipate global-scale warming trends that become more intense the more CO_2 we release, and most also anticipate somewhat wetter conditions in the northeastern United States by 2100 AD.

So what does this tell us about the future of the temperate zone? If we follow a moderate, 1,000-Gton scenario, then both the models and common sense say that most of the temperate regions should warm several degrees further by 2100 AD, especially in the higher latitudes, and that precipitation will change in a more patchy fashion. According to the combined results from all the Climate Wizard models, most of North America, northern Europe, and central Asia will become wetter overall, while much of the American Southwest, Patagonia, southern Australia, and the Mediterranean region will become drier. The South African Cape, as one might expect after learning about the likely poleward drift of winter storm tracks there, is also on the list of future drier places.

Progressively higher temperatures will reduce the duration and thickness of snow and ice cover in the northern United States, perhaps enough to make low-lying Lake Champlain totally ice-free but probably not enough to prevent the highest of the upland lakes from freezing in midwinter. The tips of the tallest Adirondack peaks might still turn white for a few months each year, and the skiing and snowmobiling industries might hang on longer here than in much of the rest of the Northeast, perhaps well into the twenty-second century and beyond. And one bothersome regional pollution problem may finally be solved when our fossil fuel consumption finally trails off. Coal-fired power plants and internal combustion engines will stop dumping so much acid rain here. Lucky us.

If Adirondack forests once more become botanical equivalents of the Blue Ridge and Smokies as a result of moderate-scale warming, they won't do it very quickly. Existing trees will take centuries to die of old age and to make room for different ones as the region slowly warms, and southern trees won't move up here any faster than their seeds can. Then as now, the most abrupt changes will probably be caused not by climate change but by forestry practices, fires, and imported pests. Already, we're losing our beeches to fungal infections as we've previously lost chestnuts and elms to foreign blights, and invasive emerald ash borers are just beginning to threaten ash trees all over North America. This also applies to animals, as well; white nose syndrome is rapidly decimating our local bat populations, yellow perch and golden shiners are driving native trout from our upland lakes, and imported zebra mussels are replacing native mussels in lowland waters.

If somewhat wetter conditions also develop here as most models suggest, then the additional moisture may or may not help to keep local wetlands and forest soils from drying out during warmer summers, depending on how the seasonal inputs and evaporation balance out; seasonal-scale model projections of precipitation in the relatively small Adirondack and Champlain regions differ too much to be reliable, though most of the models predict annual-scale wetting trends here. Warming could make meltwater surges and ice jams more frequent in our rivers during winter, but by keeping the snowpack and ice cover from building up as much it might also reduce the severity of those events in early spring. More precipitation could be a good thing for the Adirondacks in general, but it might not be so nice for cities downstream if river discharge continues to increase as well. As the ocean-linked level of New

York Harbor slowly rises around the rim of Manhattan Island, episodic but potentially destructive river flooding may also occur there if more North Country runoff events push the Hudson over its banks.

Some southern-type animals that are rarely seen in the high country today, such as opossums, may become more common here, while other rarities, like northern-type bog lemmings, may disappear. But most of our mammals, from bears and raccoons to otters and red foxes, are resilient enough to stick it out for the long haul, and all are widely distributed outside the Adirondacks. Even if lemmings and other boreal species no longer find suitable homes here, their disappearances will be local, not total, if future climatic shifts remain moderate, and their descendants may eventually return from more northerly regions after the thermal peak passes. Many of the maps of species ranges that U.S. residents consult show only blank spaces north of the national border, as if the vast climatic refuge of Canada doesn't exist. Phrases such as "extinct within New York State" are human-centered constructs, and they need not represent true extinction at all. A more species-centered approach, rather than a state- or nation-centered one, may show that bog lemmings can continue to thrive in Canada and perhaps even expand the northern limits of their range as their southern limit drifts over and beyond our park boundaries.

One often hears that global warming will cause disease-bearing insects and ticks to spread into higher temperate-zone latitudes and elevations as well. That may be true in some cases, but probably not very often in the case of mosquito-borne malaria. Claims that the tropical scourge will soon invade the United States, for example, fail to recognize that malaria is already endemic to North America. Outbreaks were common as far north as Canada during the nineteenth century, and New York City residents often suffered from it. I have an old field guide to camping in Maine, published in 1879, that warns tourists of the malarious "miasm" that was supposedly emitted by northern wetlands at night. Malaria was only recently extirpated in North America by aggressive human intervention, including the draining of breeding pools, the spraying of pesticides, the screening of windows, and improvements in health care. Similar preventive measures also drove endemic malaria from much of Europe and Scandinavia, and humans are very likely to keep it from returning to its original northerly haunts.

But what if we take the 5,000-Gton path? According to Climate Wizard, the temperate zone as a whole may warm twice as much by 2100 A.D. as it would in the moderate B1 emissions scenario, but the distribution of changes in precipitation is expected to be similar, with the Mediterranean region, the American Southwest, Patagonia, southern Australia, and the South African Cape drying, while most other temperate regions become wetter, although the magnitudes of those changes will be greater than in the milder situation.

The Adirondack Mountains are only a few million years old, so they've never been exposed to hothouse conditions as extreme as those of a PETM-like 5,000-Gton scenario. A super-greenhouse of that intensity could create conditions during the long centuries of thermal maximum that would be unlike any in local ecological history. Our highest lakes and peaks could become totally ice-free in winter, and rare alpine tundra plants such as dwarf willow and cushion-like *Diapensia*, which were originally pushed up to the tallest summits by postglacial warming, would be shoved the rest of the way upward into local extinction. In this case, the loss of boreal species from the Adirondacks might also be followed by total extinction if the Arctic heats up enough, too.

Nonetheless, mountainous regions such as the Adirondacks will have a built-in advantage over other temperate-zone areas as far-future climates warm and then cool on the long tail of the CO_2 curve. Many species will be able to migrate into new climatic settings simply by moving up or downhill rather than being forced beyond the boundaries of a wild refuge, as long as it doesn't warm too much and their other ecological needs are also met in the new locations. But the outlook isn't all rosy, either. Among the most vulnerable animals and plants will be those living at the tops of the highest mountains because they will have no place higher to retreat to. If our alpine tundra-style vegetation is lost to a 5,000-Gton super-greenhouse, it may not return to these peaks for hundreds of thousands of years.

The factor that will play the most important role in determining the fate of future wild lands will, of course, be us. We're the ones who introduce "invasive" alien species to their new territories, and people of the future will surely continue to help invaders spread into new territories whether by accident or design. And although strong, strictly enforced laws can keep forested islands like the Adirondacks afloat in a sea of development, not everyone wants to keep them that way, so there's no telling what the next century's legislative decisions might bring. Major reversals of what people can and cannot do with wild areas could easily cause more rapid and devastating changes than Anthropocene climate alone is likely to produce.

Will our descendants love the landscapes of the deep future as much as we love them now when the vegetation changes and the mountains and lakes hold less snow and ice in winter? Even the fanciest computer models can't tell us that. Seemingly wild places like the Adirondacks are already quite different from what they were in centuries past because of fires, logging, settlement, pollution, overhunting, invasive species, and disease. But most of us don't complain much about such things; we like our home territories pretty much as we first encountered them, whatever the pedigree. We can only hope that later generations feel the same way about the artificially warmer landscapes that will await them as the Anthropocene continues to unfold.

Epilogue

What we need to invent . . . are ways in which farsightedness can become a habit of the citizenry of the diverse peoples of this planet.
—Margaret Mead, Atmospheric Science Conference, North Carolina, 1975

I'm traveling from the Adirondacks to midcoast Maine with my partner, Kary, in order to attend a joint celebration of July birthdays; my father's, my mother's, and mine. As the Grand Isle ferry chugs eastward across Lake Champlain, we take a breezy bench seat together on the observation deck between blue water and blue sky to watch the gray, wave-cut bluffs of the Vermont shoreline approach. It's a familiar route for us, but my background research and writing for this book have helped to make simple journeys into living tours of the subjects discussed in these pages.

Just before I set the emergency brake and switched our car's engine off for the ferry crossing, the radio aired a short piece about how people throughout the world are becoming increasingly concerned about climate change. At the end of the broadcast Kary asked, "What would a book about the deep future have to say to those people?"

Good question. If I had to boil it all down to one sentence, I'd consider saying, "Don't panic and don't give up!" Climate change is a troubling and complex issue, but it's not going to kill us all off, either. During the last hour, Kary and I increased the total CO_2 content of our lungs and raised the mean annual temperatures of our surroundings 3°F (2°C) simply by driving downhill from Paul Smiths to Plattsburgh, where the air is both denser and warmer. But the landscape, though changed, still looks healthy and prosperous. To paraphrase a recent comment by oceanographer Wallace Broecker, we need to think, speak, and act rationally if we're going to deal successfully with the enormous environmental and social problems facing us.

But I would also hasten to add that today's carbon crisis is nevertheless a very troubling problem. The effects of our emissions on climates, oceans, and isotopes will last orders of magnitude longer than most people yet realize. They are already changing the world in important ways, and although our own species will survive those changes, many others may not.

As the ferry draws nearer to the layered cliffs, I remember telling my geology students that these rocks used to lie beneath a tropical ocean, and that some of the fossil reef deposits nearby are among the oldest in the world, even older than the fossil fuels that power our cars and ferries. During the last 450 million years, the thick plate of continental crust that now supports them has drifted far north of its former position, and what was once a thriving and colorful marine habitat is now a flat-lying tombstone with the stiff images of former inhabitants imprinted on it. The cinnamon roll spirals of ancient tropical snails are not only out of place on dry land half a hemisphere away from the equator; they don't even exist anymore except in fossil form, and neither does the former ocean basin they once lived in. If those creatures had had enough brain capacity to think millions of years into the future, what would they have thought of this extreme turn of events? Would the associated localized cooling from tropical to temperate conditions and the demise of an entire ocean have seemed like an utter catastrophe, or would vibrant Burlington, the ferry, and our pleasant cross-country jaunt have reassured them that life could still be worth living in Anthropocene times?

Farther down the road amid Vermont's lovely Green Mountains, a fuel truck pauses to scope a railroad crossing, and it makes me change my message to "Stop, Look, and Listen." A deeply historical perspective can make modern greenhouse heating seem no more outlandish than the natural PETM and Eemian warm periods of the distant past, and it can make post-whiplash cooling seem no more frightening than the long Cenozoic descent into repeated ice ages. But even so, it's still wise to stem our carbon emissions as much and as rapidly as possible. Not because it would prevent climate from changing at all; we'll still have to deal with things like the Arctic oscillation, solar cycles, and El Niño no matter what we do about carbon pollution. And not because warming is necessarily all bad, either; the formerly tropical Champlain snails certainly wouldn't think so, and preventing the next ice age is a favor to future generations.

A stronger argument for controlling our fossil fuel consumption is that a prompt switch to alternative energy sources is in our own best interest. If we don't do it soon, then we'll merely be putting it off until the eventual depletion of reserves forces our descendants to make the switch later on, and the losses of species and habitats to the climatic disturbances and ocean acidification that would follow an extreme-emissions scenario are both unethical and undesirable. If we stop our carbon pollution sooner rather than later, we can always resume it some other time if we so choose, but rushing blindly ahead now destroys options for the future. Eventually, climates will once again resemble those we know today, but by that time many of our companion species may be gone; climatic conditions are temporary, but extinction is forever.

That would be the "stop" part of my message. The "look" part would refer to learning more about how our planet operates and how our actions influence it. We still don't know which species will be able to adapt to future conditions and which ones will be lost if we take a particular emissions path. We don't even know how many species exist yet, much less how they live and interact with their surroundings. There is also much to learn about natural climate variability, ocean circulation, and the inner workings of ice sheets. Unfortunately, many of us seem to believe that the natural sciences are arcane and narrow academic specialties in comparison to other subjects such as finance, fashion, or politics, all of which are actually quite narrowly anthropocentric and, in isolation from the natural sciences that deal with the entirety of physical existence, are likely to produce short-sighted responses to environmental problems.

I ponder this depressing thought as Kary points out the rich palette of flowers lining the road. "Look at that big patch of blue chicory," she exclaims, "and those Queen Anne's lace blossoms look pretty, too." Most of the vehicles before and behind us are driven by people with other things on their minds, for whom the flanking collages of trees, shrubs, and herbs are just a featureless blur. I'm thankful to be sharing this journey with someone who sees much more than that, but I wonder how most of society can be expected to make scientifically informed decisions about long-range planetary management without an accurate sense of all that may be at stake.

Even among those of us who put specific names on wild animals and plants, a sense of ecological history is often missing. As we contemplate a warmer future in, say, the Adirondacks, it's good if we can describe an anticipated transition from sugar maple to oak and hickory forest, but it's not sufficient. What are we to make of such a change? How do we weigh the pluses and minuses in choices that will affect thousands of species and countless generations? If we automatically assume that anything other than present conditions are intolerable, then we ignore one important fact. Present conditions aren't normal, either. The Anthropocene is here already, whether we recognize it or not.

Consider those fragrant chicories and Queen Anne's laces. They brighten up our roads and meadows, but they're not native to this part of the world. They were carried here from Europe by people. Early Romans used to fry chicory in garlic, and Queen Anne ruled the British Isles in the early eighteenth century. Or what about those old feral apple trees on the forest edge? Apple pie is supposed to be as American as . . . itself, but that trademark Yankee fruit was originally imported from central Asia. In fact, the majority of our roadside and meadow plants are aliens, from purple clover and common buttercups to oxeye daisies and yarrow. To most of us they are aromatic and colorful sources of delight, and such immigrant plants have become welcome sources of food and medicine for Americans in the past. But to pro-native gardeners and the U.S. Department of Agriculture's Natural Resources Conservation Service, these are best described as "noxious weeds." Viewed from a historical perspective, their arrival on the continent is a source of dismay to those who favor North American species over others.

The same is true of many of our animals, too. The dark starlings poking about on that lawn are human-borne invaders and so, too, are the honeybees that pollinate its "weeds" and the earthworms that burrow beneath it, aerate the soil, and possibly bait a kid's fish-hook. The ancestors of the brown trout in that Vermont stream were hauled across the Atlantic Ocean from Germany, and the rainbow trout hail from the West Coast. Even the relative abundances of local creatures are skewed by human activity; white-tailed deer and coyotes now far outnumber moose and wolves in the Northeast, reflecting the legacy of our hunting, farming, and forestry practices.

The Anthropocene has been under way for so long that most of us don't even notice it, having grown up with it as the normal state of things. As we think ahead to the even longer stretches of human history that lie before us, it can also be interesting to ask what our times might seem like from the point of view of those who lived in earlier stages of this ever-changing human epoch. If we imagine ourselves living in New England in the 1700s, and we then imagine looking from that vantage point into the future, does the rural landscape as we now know it seem unattractive because it no longer looks like it did when Queen Anne ruled the colonies? Would a longer step back to the 1500s make us protest the Spanish introduction of culture-changing horses to this continent? And would a step even farther back in time make us mourn the arrival of the first Stone Age humans who would slaughter the American mammoths and mastodons and thus leave our landscapes unnaturally silent—but much safer to walk in?

I expect not, at least not for most of us. We love our Anthropocene world, artificially altered though it is by our very presence in it. Perhaps the same will be true of those who will live in the artificially altered versions of the future world that follow this one. How can we tell which changes are truly "bad" for all or most of the parties involved, and which ones might simply come to seem normal or even desirable later on?

Crossing into New Hampshire on a bridge over the headwaters of the Connecticut River, I revisit my choice of summary phrases again and realize that the "listen" component of it might present the greatest hurdle to becoming "responsible ancestors" in the next chapters of history.

This is the first period of Earth's existence in which a single species *consciously* occupies and manipulates the entire planet. One of our most transformative evolutionary steps as a social species was the sharing of knowledge; you might not know which roots to dig up during a famine but if Grandmother does, her childhood memories can save a whole village when hard times strike. But now we're being challenged to take another giant step, from shared knowledge to shared responsible action on a global scale.

In this new Age of Humans, our very thoughts and desires have become powerful environmental forces in their own rights, and how we think and act can be as important to millions of other humans (and other species) as to ourselves. The better we know and respect each other as people, the more we're likely to learn from one another, the more likely we are to understand each other's needs and goals, and the more likely we are to cooperate effectively for our mutual benefit. Greenhouse pollution problems will not be solved piecemeal, and there is also no way to avoid making a collective choice one way or the other. We'll either decide to solve them together as a self-aware global community or we'll decide to suffer through them together as a disjointed mob of individuals.

From this point of view it's clearly worth putting the brakes on until we've had time to figure out where we are and where we want to go from here. There are many difficult questions to resolve, some of which may be painful. Any choice we make will probably benefit some people while harming others. Maybe a global hothouse could be more user-friendly for tropical nations if the low latitudes become wetter and for circumpolar nations if the Arctic becomes more habitable and navigable. On the other hand, those who live on the edges of expanding deserts would probably rather not see such things come to pass, and coastal peoples would surely rather not have to deal with sea-level instability and ocean acidification. The smartest and most ethically sound solution is to pause, listen to one another carefully and respectfully, and then try to move ahead as a single species on our singular planet.

This, of course, is a challenge considering the practical limits of altruism, which many of us still limit to immediate family and friends as our ancestors did in smaller, less interconnected versions of the world. And the divisive gamesmanship and spin of modern politics and media are major obstacles now, as well. Convincing a majority to support any particular climate control strategy isn't going to be easy, especially if it involves international laws that seem to threaten national sovereignty. But, to paraphrase John Lennon, even simply imagining the possibility of humankind making sound decisions together can be a good way to start. Creative consciousness-raising events such as the International Day of Climate Action organized by 350.org can be powerful sources of inspiration in that regard even for those who doubt that 350 ppm CO_2 is a realistically attainable target in the near-term future.

We pull over at a rest stop for a stretch break, and I nose our car into an open parking space beside a battered pickup truck whose back end is plastered with a diverse collection of bumper stickers. "Save the Whales," says one. "Save the Planet," says another. But a sticker on another truck beside that one seems to be trying to counter them with a sarcastic "Save the Humans." Do we always have to split so reflexively into opposing sides when discussing important issues such as global climate change?

If we continue to burn fossil fuels at current rates, then those who enjoy or profit from their use may benefit, but the resultant emissions will afflict future societies and species with thousands of years of artificial climatic change and the multitude of cultural and environmental problems that would come with it. On the other hand, if we reform our combustive way of life too quickly then many of us may suffer hardships in the near term, and the citizens of 130,000 AD may also be doomed to endure an ice age. It seems like a Faustian bargain either way when we consider the full span of the Anthropocene epoch that lies ahead.

But maybe there's a middle route. If we do manage to follow a moderate-emissions path, then we'll probably be leaving most of our coal reserves where they lie and running our future civilizations on other energy sources. Environmental damage during the next several centuries will be held to a minimum, some societies might benefit from a partial and temporary opening of the Arctic Ocean, and the next ice age of 50,000 AD will be held at bay. This could also produce a longer-term benefit, as well, by leaving lots of coal already sequestered in the ground for later. By saving most of our fossil carbon in a safe, solid, reasonably accessible form, we would bequeath it to later generations for possible use, not necessarily as a fuel but rather as a simple, cost-effective tool for climate control.

Even if residents of 130,000 AD lack complex modern-style technologies, they could protect themselves from the ice age that is due to start then if they understand what is going on in the climate system and remember that CO_2 is a greenhouse gas. The only technology that would be required to mobilize it would be fire, one of humankind's most simple yet powerful tools. And by setting reasonable quantities of buried coal deposits alight, future climate modifiers would not only be able to keep the global thermostat at whatever they deem to be an optimal setting at a crucial time of cyclic cooling, they could also harness the heat and light that coal combustion produces. Releasing CO_2 exhaust in this manner would still dump artificially generated fumes into the air, but doing so in a responsibly controlled manner rather than burning it wholesale would minimize its harmful side effects. And even though coal supplies would still eventually run out in the very far future, our descendants could still keep many more ice ages at bay than would otherwise be possible with a single massive emission pulse in our own current century.

What would it cost to "save the carbon" today? It would require us to find alternative energy sources, and fast. But we need to do that anyway. We're already near the limits of economically viable petroleum production, and the decline of cheap oil will have swift and severe consequences for those who will inherit the full measure of that problem. If and when the prices and availability of petroleum-based fuels, fertilizers, plastics, pharmaceuticals, cosmetics, synthetic fabrics, and even roadway pavement go haywire, the scale of human suffering could outstrip anything in the works for us from climate. It is that horribly dangerous and fast-approaching situation that makes the need for a switch to nonfossil fuels a no-brainer of the first order.

From a full Anthropocene-scale perspective, coal is both too valuable and too environmentally damaging to burn indiscriminately. Its highest use is as a long-term climate protection device, not just cheap and dirty furnace food. Running power plants on it is like burning your house down around you because it's cold outside. It's like cutting a square of fabric out of the floor of your life raft to patch a hole in your trousers. In short, it's . . . well, it's kind of stupid.

As we pull back onto the road after pondering this idea together, Kary chuckles and begins to sketch the design of a bumper sticker for our own car. It's black with white block letters and it reads "Save the Carbon." Save it for later, for humankind to deal with wisely both now and in the future. If you see us driving down the road with that sticker plastered onto our back bumper, give us a friendly honk if you agree with it.

Route 2 leads us through the White Mountains and into western Maine. While passing through one in a series of pulp-scented paper mill towns, I slow down to watch the frothing Androscoggin River churn over a series of power dams. Kary asks the question that I'm thinking. "What are we going to run the world on without cheap fossil fuels?"

The carbon crisis that we face today involves more than pollution problems; it also includes the struggle to find enough affordable, sustainable nonfossil fuels to run our societies on. My guess is that a combination of alternative energy sources, rather than any single one, will be needed to wean us of our carbon dependence. And I hope that it happens quickly, because the other options are lousy. It would be perilous to continue with business as usual and thereby take the extreme-emissions path. But artificially sequestering enough carbon to return the atmosphere rapidly to preindustrial conditions may be so costly as to be unrealistic, considering human nature. Who will really be willing to pay for it when most of the benefits go to future generations and when some parties inevitably refuse to pitch in? And artificially raising the price of fossil fuels in order to force us to reduce consumption would essentially make energy too expensive for many of us to use, an unthinkable hardship to impose on anybody who has ever struggled to make ends meet. Our best hope is to encourage the development of new sources of energy, and I'm cautiously optimistic about it for two reasons: We have nuclear power in our back pocket if we really need it as a last resort, though present forms of it come with extremely serious unsolved problems of long-term safety and waste disposal. And we have billions of creative, interconnected people to tap for new and better ideas.

To me, one of the most exciting prospects for a new energy source is hydrogen fuel. Not just the kind that is being derided in some circles for being unrealistic, though. The version that is most often under discussion now has us splitting water molecules with electricity that must be generated by yet another power source, and to critics it seems like a waste of electrons that could be harnessed more efficiently for other purposes. In places like Iceland, where hydro and geothermal energy are naturally abundant and cheap, electrically generated hydrogen is already making notable inroads into the economy, and Iceland has recently built the first of a planned fleet of hydrogen-powered ships. But my favorite generative source is different, and it could be used almost anywhere on Earth. It's photosynthesis.

Plants, algae, and bacteria have been splitting water molecules into their oxygen and hydrogen subunits for hundreds of millions of years, and they get their requisite energy from sunlight. The molecular details of how they do it and how we could mimic them through nano-technology are now being worked out by botanists and molecular engi-neers around the world, including teams at Australia's Monash University, the Swiss Federal Institute of Technology, and Penn State and Rutgers Universities in the United States. If all goes according to plan, synthetic solar water-splitters that resemble or even outperform natural photo-systems might some day give us the mother of all green—or should we say, blue—fuels. The hydrogen source would be water and the main product of its combustion would be water, too; that's how this lightest of all gases got its name in the first place—from hydrogenesis, the creation of water.

In my daydreams about this up-and-coming energy resource, I imagine rooftops rustling with leafy collector foliage, lawns converted to hydrogen gardens, and green cars slathered with photosynthetic paint. More likely, though, most solar hydrogen would be produced by central-ized commercial operations and distributed through pipelines and pres-surized containers, and it would be supplemented by electrically produced hydrogen, power dams, windmills, and other noncarbon energy sources. But no matter how new energy technology really develops in the years to come, let's hope that it becomes commercially viable much sooner than later.

The sun is sinking closer to the crest of the White Mountains be-hind us as Kary takes the wheel, and the road breeze that flows over my right arm and curls in through the open passenger-side window feels cooler than it did a few hours ago. The temperature change draws my at-tention to the air current, and my imagination populates it with mole-cules that I know are there but are too small to see. The flow becomes a smooth stream of particles sliding like fine powder over my skin. With my eyes still fixed on the road ahead, a lightly meditative trance induced by long hours of driving paints colors on the invisible particles, and in my mind the wind takes on the pale hue of blowing sand.

Most of the grains are light brown—that's nitrogen. Nobody pays it much attention except as an inert lung-filler or as the preferred diet of a few microbes. About a fifth of them are white—that's free oxygen. It wouldn't be here if photosynthesis had never evolved to pump it out as waste, so by rights it ought to be called air pollution. But if photosynthesis had never existed, then neither would we, so we won't be faulted for being thankful that plenty of oxygen is here for us to breathe. And less than one in a hundred of the little grains are gray—that's CO_2, a black carbon dot pinned between two white oxygen atoms.

It strikes me as odd that I, an investigator of climate change who has apparently been somewhat "blinded by science," have never made the gut-level connection between global warming and the touch of wind on my skin until this moment. The gases that are warming the world, acidifying the oceans, and changing the isotopic compositions of our bodies are not just words on paper or formulae on some professor's chalkboard. They are filling your lungs and bumping into your face as you sit here reading my words and vibrating in a professor's throat as she delivers a lecture. They flood the seemingly empty spaces between our eyes and the horizon, and between our hats and the clouds. As the diffuse substance of wind, they caress us in summer and freeze us in winter. And by the end of this century, there may be twice as many of them in the transparent sea of air as there were in 1750. Artificial greenhouse gases are as real and present in our daily lives as the Anthropocene is; as Bill McKibben argued in *The End of Nature,* they have already erased the last vestiges of untouched wildness anywhere on the planet. They're right here among us and inside of us, they're increasing in abundance, and they're determining the climatic future of the world. If they were large enough to be seen with the unaided eye, it would be impossible for anyone to ignore them.

A thick bank of chilling mist engulfs us as we pull into the waterfront town of Belfast, just a few miles away from my father's house on the edge of Penobscot Bay. The early-evening air is calm and moist, and a foghorn hoots a low, resonant warning in the distance. Herring gulls shuffle aside to make room for our car in the parking lot overlooking the harbor, or what can be seen of it in this thick whiteness. We step out into a delicious tangle of smells that have been amplified by the humidity; the tang of salt, the sweetness of mudflats and seaweed, and the mouthwatering aromas of steamed lobster and melted butter leaking from the door of a nearby restaurant. First things first, though: we head for the water's edge before following the scent trail to dinner.

Brown, nut-sized periwinkle snails are grazing on soft algal films that coat the wet granite blocks on either side of the boat launch ramp. These animals, now as common as pebbles on this rocky coast, are accidental invaders who emerged from the watery holds of European ships during earlier centuries along with the little green crabs that scuttle among tufts of rubbery brown rockweed just below the waterline. And like me and everyone else who lives in the Americas, even the rocky coast itself is an immigrant of sorts. This section of eastern Maine was thrust up and welded onto the continent when a landmass that geologists call Avalon collided with proto–North America. It happened nearly half a billion years ago amid a long series of collisions that built the White Mountains on today's route and crushed the ocean that the Champlain fossil reef once thrived in. Since its arrival on this shore, the remnant strip of Avalon has been roasted by volcanoes, scoured by ice sheets, and alternately stranded and flooded by glacial and interglacial sea-level changes.

Why fear change when we live in such an inconstant world? I sometimes wonder if it's not global warming that worries us so much as change of any kind. Take the rising trends of temperature or sea level from this century and flip them over their respective whiplash peaks into the reversed modes that will dominate the long tail of the CO_2 curve, and they can seem just as worrisome either way. The retreat of Arctic sea ice may be dreadful to an Inuit seal hunter today, but the eventual refreezing of the polar ocean in the distant future could be equally troubling to northern fisherfolk who will live on the far side of the Anthropocene carbon peak.

There may be good practical reasons, dating back to the earliest days of more tenuous human existence, for disliking change. In those times it might have brought a risk of starvation, or death by predators or foes, or dangerous exposure to the elements. But today we face another situation altogether. By succeeding so spectacularly, crowding into every imaginable habitat and weaving ever more complex social and economic networks, our species is now pressed so tightly against the physical limitations of life on a finite planet that almost any kind of environmental disturbance is potentially disruptive. Because we live everywhere, someone is bound to suffer somewhere regardless of whether it warms or cools or wets or dries. Our boat has been overloaded with gear and passengers, so it's perfectly reasonable to fret about the wind and waves. But I can't help thinking that it's not so much the weather that is ultimately to blame as it is our own behavior.

The edge of the sea is mostly smooth and still beneath its heavy blanket of fog. But from time to time an isolated swell rolls into view and pushes the edge a little farther up the boat ramp. Perhaps it's an advance message from an approaching storm, a hint of rougher weather ahead, and it makes me think of the early signs of Anthropocene climate change and ocean acidification that are already stirring many of the more alert among us into action.

I can understand why some activists feel justified in nudging the human herd in better directions by hook or by crook. But although I do hope that we control our pollution and stop driving so many species to extinction, I would rather have us do it on purpose rather than be tricked into it.

That's where science comes in. In a media-saturated world where public opinions are easily swayed by team loyalty, marketing strategies, and short-term self-interest, science stands apart as a rare source of relatively impartial, self-correcting information. The strict rules of scientific investigation favor well-supported ideas over weak ones, and the international peer-review system is a firewall of checks and balances that provides an additional line of defense against sloppy or slanted thinking. Good science provides a universal knowledge base that is unusually free of ulterior motives and spin. As that knowledge evolves with the input of new data and ideas, you may sometimes be frustrated by having it change your mind from time to time; I'm sure, for example, that some of the things that I've written on these pages might eventually require updating in the face of newly acquired information. But at least that challenge to your current worldview is not the result of someone trying to manipulate you for some hidden, possibly nefarious purpose.

This is why aggressive activist stances among prominent scientists make me nervous. Most scientists try to stick closely to the facts, following the suggestion of ecologist Erle Ellis and geologist Peter Haff in a recent issue of *Eos* to "remain free of intentional distortion or personal bias" when giving advice in public. But I also know that at least one well-known figure in the climate community has purposely exaggerated the dangers of global warming in public presentations, because he told me so at a conference. His justification was this: "If people aren't scared, they won't pay attention."

Environmental professionals today are being called upon to use their scientific credentials as intellectual weapons, and some researchers are tempted to play up the fear factor and downplay their uncertainties in order to shape and hold public attention. But there's an important difference between informing and promoting or selling, and once you step over that line you have left the fortress of objectivity to become just another mercenary whacking away at people. Worse still, you also risk dragging the good name of impartial science down with you into the muck of the battlefield. I believe that scientists are most valuable to society when they're seen to stand a little apart from the fad-driven mainstream in order to raise or answer questions as impartially and insightfully as possible. It may be a lonely way to live at times, but I believe that it's worth it; if people stop listening to scientists because they seem to take sides unfairly, then we're all in trouble.

We have a lot of work to do if we're going to deal with our carbon crisis in a responsible manner, and to do it well we're going to need to learn a lot more about each other and about the environments we live in, not just as they are now but also in the context of history and the foreseeable future. Hopefully, this book can help a bit in that regard. In any case, I'm confident that we have the smarts, the heart, and the time necessary to succeed if we set our minds to it.

And that leads to the most important take-home message of all. Whatever we decide, we are the ones who will chart the course of this human-centered age and determine the fate of the ecosystems and species that will share the future Earth with us. In a literal sense that can be understood with or without the support of a religious tradition, we are participating in a new creation, the genesis of a world over which we hold an ever-increasing measure of dominion.

We are not infallible gods, of course, and our role in this remaking of the world is a heavy burden of responsibility to bear, but it is also liberating to recognize the surprising degree of influence that we now have on our home planet, as well as the age-old influences that it still has on us. Finding the right balance between power and responsibility will be our primary task as we struggle to mature further as a wise species, hopefully to more fully merit the pretentious name that we have given ourselves, *Homo sapiens.*

For better or for worse, we are both the products and the creators of this remarkable new Age of Humans, and we will be the ones to decide the direction it takes from here on into the deep future.

Welcome, everyone, to the Anthropocene.

References

Prologue

Archer, D. 2005. "The Fate of Fossil Fuel CO_2 in Geologic Time." *Journal of Geophysical Research* 110: C09805, doi:10.1029/2004/C002625.

——— and V. Brovkin. 2008. "The Millennial Atmospheric Lifetime of Anthropogenic CO_2." *Climatic Change* 90: 283–297.

——— and A. Ganopolski. 2005. "A Movable Trigger: Fossil Fuel CO_2 and the Onset of the Next Glaciation." *Geochemistry, Geophysics, Geosystems* 6: Q05003, doi:10.1029/2004GC000891.

Crutzen, P. 2002. "The Geology of Mankind." *Nature* 415: 23.

———. 2006. *Earth System Science in the Anthropocene*. Berlin: Springer.

Crutzen, P., and E. F. Stoermer. 2000. "The 'Anthropocene.'" *Global Change Newsletter* 41: 12–13.

Gill, J. L., J. W. Williams, S. T. Jackson, K. B. Lininger, and G. S. Robinson. 2009. "Pleistocene Megafaunal Collapse, Novel Plant Communities, and Enhanced Fire Regimes in North America." *Science* 326: 1100–1103.

Kump, L. R. 2008. "The Rise of Atmospheric Oxygen." *Nature* 451: 277–278.

Meehl, G. A., et al. 2007. "Global Climate Projections." In: *Climate Change 2007: The Physical Science Basis. Contribution of Working Group I to the Fourth Assessment Report of the Intergovernmental Panel on Climate Change*, S. Solomon et al., eds. Cambridge, UK: Cambridge University Press.

Ruddiman, W. F. 2005. *Plows, Plagues, and Petroleum: How Humans Took Control of Climate*. Princeton, NJ: Princeton University Press.

1. Stopping the Ice

Archer, D., and A. Ganopolski. 2005. "A Movable Trigger: Fossil Fuel CO_2 and the Onset of the Next Glaciation." *Geochemistry, Geophysics, Geosystems* 6: Q05003, doi:1029/2004GC000891.

Berger, A., and M. F. Loutre. 2002. "An Exceptionally Long Interglacial Ahead?" *Science* 297: 1287–1288.

Broecker, W. S. 1999. "What If the Conveyor Were to Shut Down? Reflections on a Possible Outcome of the Great Global Experiment." *GSA Today* 9: 1–7.

———. 2006. "Abrupt Climate Change Revisited." *Global and Planetary Change* 54: 211–215.

———. 2006. "Was the Younger Dryas Triggered by a Flood?" *Science* 312: 1146–1148.

———. 2009. "Future Global Warming Scenarios." *Science* 304: 388.

Bryden, H. L., H. R. Longworth, and S. A. Cunningham. 2005. "Slowing of the Atlantic Meridional Overturning Circulation at 25° N." *Nature* 438: 655–657.

Cochelin, A.-S., L. A. Mysak, and Z. Wang. 2006. "Simulation of Long-Term Future Climate Changes with the Green Mcgill Paleoclimate Model: The Next Glacial Inception." *Climatic Change*, doi:10.1007/S10584-006-9099-1.

Crucifix, M., and A. Berger. 2006. "How Long Will Our Interglacial Be?" *Eos* 87: 352–353.

Drysdale, R., J. C. Hellstrom, G. Zanchetta, A. E. Fallick, M. F. Sánchez Goñi, I. Couchoud, J. McDonald, R. Maas, G. Lohmann, and I. Isola. 2009. "Evidence for Obliquity Forcing of Glacial Termination II." *Science* 325: 1527–1531.

Hays, J. D., J. Imbrie, and N. J. Shackleton. 1976. "Variations in the Earth's Orbit: Pacemaker of the Ice Ages." *Science* 194: 1121–1132.

Kerr, R. 2006. "False Alarm: Atlantic Conveyor Belt Hasn't Slowed Down After All." *Science* 314: 1064.

Kukla, G. J., R. K. Matthews, and J. M. Mitchell. 1972. "Present Interglacial: How and When Will It End?" *Quaternary Research* 2: 261–269.

Meehl, G. A., W. M. Washington, W. D. Collins, J. M. Arblaster, A. Hu, L. E. Buja, W. G. Strand, and H. Teng. 2005. "How Much More Global Warming and Sea Level Rise?" *Science* 307: 1769–1772.

Pollard, D., and R. M. Deconto. 2009. "Modelling West Antarctic Ice Sheet Growth and Collapse Through the Past 5 Million Years." *Nature* 458: 329–332.

Rahmstorf, S. 2003. "The Current Climate." *Nature* 421: 699.

Raymo, M. E., and P. Huybers. 2008. "Unlocking the Mysteries of the Ice Ages." *Nature* 451: 284–285.

Schiermeier, Q. 2007. "Ocean Circulation Noisy, Not Stalling." *Nature* 448: 844–845.

Schwartz, P., and D. Randall. 2003. "An Abrupt Climate Change Scenario and Its Implications for United States National Security." http://www.mindfully .org/air/2003/pentagon-climate-changeloct03.htm.

Short, D. A., and J. G. Mengel. 1986. "Tropical Climate Phase Lags and Earth's Precession Cycle." *Nature* 323: 48–50.

Sirocko, F., K. Seelos, K. Schaber, B. Rein, F. Dreher, M. Diehl, R. Lehne, K. Jäger, M. Krbetscek, and D. Degering. 2005. "A Late Eemian Aridity Pulse in Central Europe During the Last Glacial Inception." *Nature* 436: 833–836.

Sternberg, J. 2006. "Preventing Another Ice Age." *Eos* 87: 539, 542.

Toggweiler, J. R., and J. Russell. 2008. "Ocean Circulation in a Warming Climate." *Nature* 451: 286–288.

Vernekar, A. D. 1972. *Long-Period Global Variations of Incoming Solar Radiation*. *Meteorological Monographs* 12. Boston: American Meteorological Society.

Weaver, A. J., and C. Hillaire-Marcel. 2004. "Global Warming and the Next Ice Age." *Science* 304: 400–402.

Wunsch, C. 2002. "What Is the Thermohaline Circulation?" *Science* 298: 1179–1181.

2. Future Carbon

Allen, M. R., D. J. Frame, C. Huntingford, C. D. Jones, J. A. Lowe, M. Meinshausen, and N. Meinshausen. 2009. "Warming Caused by Cumulative Carbon Emissions Towards the Trillionth Tonne." *Nature* 458: 1163–1166.

Archer, D. 2005. "The Fate of Fossil Fuel CO_2 in Geologic Time." *Journal of Geophysical Research* 110: C09805, doi:10.1029/2004/C002625.

———. 2007. "Methane Hydrate Stability and Anthropogenic Climate Change." *Biogeosciences* 4: 521–544.

———. 2008. *The Long Thaw: How Humans Are Changing the Next 100,000 Years of Earth's Climate*. Princeton, NJ: Princeton University Press.

——— and V. Brovkin. 2008. "The Millennial Atmospheric Lifetime of Anthropogenic CO_2." *Climatic Change* 90: 283–297.

——— and A. Ganopolski. 2005. "A Movable Trigger: Fossil Fuel CO_2 and the Onset of the Next Glaciation." *Geochemistry, Geophysics, Geosystems* 6: Q05003, doi:10.1029/2004GC000891.

————et al. 2009. "Atmospheric Lifetime of Fossil Fuel Carbon Dioxide." *Annual Review of Earth and Planetary Sciences* 37: 117–134.

Berner, B. A., A. C. Lasaga, and R. M. Garrels. 1983. "The Carbonate-Silicate Geochemical Cycle and Its Effect on Atmospheric Carbon Dioxide over the Past 100 Million Years." *American Journal of Science* 283: 641–683.

Caldeira, K. 1995. "Long-Term Control of Atmospheric Carbon Dioxide: Low-Temperature Sea-Floor Alteration or Terrestrial Silicate-Rock Weathering." *American Journal of Science* 295: 1077–1114.

———— and G. H. Rau. 2000. "Accelerating Carbonate Dissolution to Sequester Carbon in the Ocean: Geochemical Implications." *Geophysical Research Letters* 27: 225–228.

———— and M. E. Wickett. 2005. "Ocean Model Predictions of Chemistry Changes from Carbon Dioxide Emissions to the Atmosphere and Oceans." *Journal of Geophysical Research: Oceans* 110: (C9).

Canadell, J. G., C. Le Quéré, M. R. Raupach, C. B. Field, E. T. Buitenhuis, P. Ciais, T. J. Conway, N. P. Gillett, R. A. Houghton, and G. Marland. 2007. "Contributions to Accelerating Atmospheric CO_2 Growth from Economic Activity, Carbon Intensity, and Efficiency of Natural Sinks." *Proceedings of the National Academy of Science, USA* 104: 10288–10293.

Crutzen, P. 2002. "The Geology of Mankind." *Nature* 415: 23.

————. 2006. *Earth System Science in the Anthropocene*. Berlin: Springer.

———— and J. W. Birks. 1982. "The Atmosphere After a Nuclear War. Twilight at Noon." *Ambio* 11: 114–125.

———— and E. F. Stoermer. 2000. "The 'Anthropocene.'" *Global Change Newsletter* 41: 12–13.

Eby, M., K. Zickfeld, A. Montenegro, D. Archer, K. J. Meissner, and A. J. Weaver. 2009. "Lifetime of Anthropogenic Climate Change: Millennial Time Scales of Potential CO_2 and Surface Temperature Perturbations." *Journal of Climate* 22: 2501–2511.

Fowler, C. M. R., C. J. Ebinger, and C. J. Hawkesworth, eds. *The Early Earth: Physical, Chemical and Biological Development*. London: Geological Society Special Publication 199, 259–274.

Gathorne-Hardy, F. J., and W. E. H. Harcourt-Smith. 2003. "The Super-Eruption of Toba: Did It Cause a Human Bottleneck?" *Journal of Human Evolution* 45: 227–230.

Goodwin, P., R. G. Williams, M. J. Follows, and S. Dutkeiwicz. 2007. "The Ocean-Atmosphere Partitioning of Anthropogenic Carbon Dioxide on Centennial Timescales." *Global Biogeochemical Cycles* 21: GB1014, doi:10.1029/2006GB002810.

Hansen, J. E., et al. 2005. "Earth's Energy Imbalance: Confirmation and Implications." *Science* 308: 1431–1435.

IPCC. 2007. "Summary for Policymakers." In: *Climate Change 2007: Impacts, Adaptation and Vulnerability. Contribution of Working Group II to the Fourth Assessment Report of the Intergovernmental Panel on Climate Change*, M. L. Parry et al., eds. Cambridge, UK: Cambridge University Press.

Jackson, S. 2007. "Looking Forward from the Past: History, Ecology, and Conservation." *Frontiers in Ecology and the Environment* 5: 455.

Kump, L. R. 2008. "The Rise of Atmospheric Oxygen." *Nature* 451: 277–278.

Lenton, T. M., and C. Britton. 2006. "Enhanced Carbonate and Silicate Weathering Accelerates Recovery from Fossil Fuel CO_2 Perturbations." *Global Biogeochemical Cycles* 20: GB3009, doi:10.1029/2005GB002678.

——— M. S. Williamson, N. R. Edwards, R. Marsh, A. R. Price, A. J. Ridgwell, J. G. Shepherd, S. J. Cox, and the GENIE Team. 2006. "Millennial Timescale Carbon Cycle and Climate Change in an Efficient Earth System Model." *Climate Dynamics* 26: 687–711.

Meehl, G. A., et al. 2005. "How Much More Global Warming and Sea Level Rise?" *Science* 307: 1769–1772.

Meissner, K. J., M. Eby, A. J. Weaver, and O. A. Saenko. 2007. "CO_2 Threshold for Millennial-Scale Oscillations in the Climate System: Implications for Global Warming Scenarios." *Climate Dynamics*, doi:10.1007/S00382-007-0279-0.

Monastersky, R. 2009. "A Burden Beyond Bearing." *Nature* 458: 1091–1094.

Montenegro, A., V. Brovkin, M. Eby, D. Archer, and A. J. Weaver. 2007. "Long-Term Fate of Anthropogenic Carbon." *Geophysical Research Letters* 34: L19707, doi:1029/2007GL030905.

Parry, M., J. Lowe, and C. Hanson. 2009. "Overshoot, Adapt, and Recover." *Nature* 458: 1102–1103.

Ridgwell, A., and J. C. Hargreaves. 2007. "Regulation of Atmospheric CO_2 by Deep-Sea Sediments in an Earth System Model." *Global Biogeochemical Cycles* 21: GB2008 doi:10.1029./2006GB002764.

Royal Society. 2005. "Ocean Acidification Due to Increasing Atmospheric Carbon Dioxide." Royal Society Policy Document 12/05. http://royalsociety.org/Ocean-acidification-due-to-increasing-atmospheric-carbon-dioxide/.

Ruddiman, W. F. 2005. *Plows, Plagues, and Petroleum: How Humans Took Control of Climate*. Princeton, NJ: Princeton University Press.

Schneider, S. H., and J. Lane. 2006. "An Overview of 'Dangerous' Climate Change." In: *Avoiding Dangerous Climate Change*, H. J. Schellnhuber, W. Cramer, N. Nakicenovic, eds. Cambridge, UK: Cambridge University Press.

Schmittner, A., A. Oschlies, H. D. Matthews, and E. D. Galbraith. 2008. "Future Changes in Climate, Ocean Circulation, Ecosystems, and Biogeochemical Cycling Simulated for a Business-As-Usual CO_2 Emission Scenario Until Year 4000 AD." *Global Biogeochemical Cycles* 22: GB1013, doi:10.1029/2007GB002953.

Solomon, S., G.-K. Plattner, R. Knutti, and P. Friedlingstein. 2009. "Irreversible Climate Change Due to Carbon Dioxide Emissions." *Proceedings of the National Academy of Sciences* 106: 1704–1709.

Meehl, G. A., et al. 2007. "Global Climate Projections." In: *Climate Change 2007: The Physical Science Basis. Contribution of Working Group I to the Fourth Assessment Report of the Intergovernmental Panel on Climate Change*, S. Solomon et al., eds. Cambridge, UK: Cambridge University Press.

Stager, J. C. 1987. "Silent Death from Cameroon's Killer Lake." *National Geographic* (September): 404–420.

Stockstad, E. 2004. "Defrosting the Carbon Freezer of the North." *Science* 304: 1618–1620.

Tans, P. P., and P. S. Bakwin. 1995. "Climate Change and Carbon Dioxide Forever." *Ambio* 24: 376–378.

Thomas, B. C., et al. 2005. "Terrestrial Ozone Depletion Due to a Milky Way Gamma-Ray Burst." *Astrophysical Journal* 622: L153–L156.

Thorsett, S. 1995. "Terrestrial Implications of Cosmological Gamma-Ray Bursts." *Astrophysical Journal* 444: L53–L55.

Tyrrell, T., J. G. Shepherd, and S. Castle. 2007. "The Long-Term Legacy of Fossil Fuels." *Tellus* 59: 664–672.

Wigley, T. M. L. 2005. "The Climate Change Commitment." *Science* 307: 1766–1769.

Zachos, J. C., G. R. Dickens, and R. E. Zeebe. 2008. "An Early Cenozoic Perspective on Greenhouse Warming and Carbon-Cycle Dynamics." *Nature* 451: 279–283.

3. The Last Great Thaw

Balter, M. 2009. "Early Start for Human Art? Ochre May Revise Timeline." *Science* 323: 569.

Berger, A., and M. F. Loutre. 2002. "An Exceptionally Long Interglacial Ahead?" *Science* 297: 1287–1288.

Blanchon, P., A. Eisenhauer, J. Fietzke, and V. Liebetrau. 2009. "Rapid Sea-Level Rise and Reef Back-Stepping at the Close of the Last Interglacial Highstand." *Nature* 458: 881–884.

Bosch, J. H. A., P. Cleveringa, and Z. T. Meijer. 2000. "The Eemian Stage in the Netherlands: History, Character and New Research." *Netherlands Journal of Geosciences* 79: 135–145.

Bowler, J. M., K.-H. Wyrwoll, and Y. Lu. 2001. "Variations of the Northwest Australian Summer Monsoon over the Last 300,000 Years: The Paleohydrological Record of the Gregory (Mulan) Lakes System." *Quaternary International* 83–85: 63–80.

Brewer, S., J. Guiot, M. F. Sánchez-Goñi, and S. Klotz. 2008. "The Climate in Europe During the Eemian: A Multi-Method Approach Using Pollen Data." *Quaternary Science Reviews* 27: 2303–2315.

Brigham-Grette, J., and D. M. Hopkins. 1995. "Emergent Marine Record and Paleoclimate of the Last Interglaciation Along the Northwest Alaskan Coast." *Quaternary Research* 43: 159–173.

CAPE–Last Interglacial Project Members. 2006. "Last Interglacial Arctic Warmth Confirms Polar Amplification of Climate Change." *Quaternary Science Reviews* 25: 1383–1400.

Ceulemans, R., L. Van Praet, and X. N. Jiang. 2006. "Effects of CO_2 Enrichment, Leaf Position and Clone on Stomatal Index and Epidermal Cell Density in Poplar (*Populus*)." *New Phytologist* 131: 99–107.

Chen, F. H., M. R. Qiang, Z. D. Feng, H. B. Wang, and J. Bloemendal. 2003. "Stable East Asian Monsoon Climate During the Last Interglacial (Eemian) Indicated by Paleosol S1 in the Western Part of the Chinese Loess Plateau." *Global and Planetary Change* 36: 171–179.

Clark, P. U., and P. Huybers. 2009. "Interglacial and Future Sea Level." *Nature* 462: 856–857.

Cuffey, K. M., and S. J. Marshall. 2000. "Substantial Contribution to Sea-Level Rise During the Last Interglacial from the Greenland Ice Sheet." *Nature* 404: 591–594.

Dansgaard, W., and J.-C. Duplessy. 2008. "The Eemian Interglacial and Its Termination." *Boreas* 10: 219–228.

Demenocal, P., J. Adkins, J. Ortiz, and T. Guilderson. 2000. "Millennial-Scale Sea-Surface Temperature Variability During the Last Interglacial and Its Abrupt Termination." *Eos* 81: F675. American Geophysical Union, Fall Meeting Supplement.

Drysdale, R. N., J. C. Hellstrom, G. Zanchetta, A. E. Fallick, M. F. Sánchez Goñi, I. Couchoud, J. McDonald, R. Mass, G. Lohmann, and I. Isola. 2009. "Evidence for Obliquity Forcing of Glacial Termination II." *Science* 325: 1527–1531.

EPICA Community Members. 2004. "Eight Glacial Cycles from an Antarctic Ice Core." *Nature* 429: 623–628.

Froese, D. G., J. A. Westgate, A. V. Reyes, R. J. Enkin, and S. J. Preece. 2008. "Ancient Permafrost and a Future, Warmer Arctic." *Science* 321: 1648.

Gaudzinski, S. 2004. "A Matter of High Resolution? The Eemian Interglacial (OIS 5e) in North-Central Europe and Middle Palaeolithic Subsistence." *International Journal of Osteoarcheology* 14: 201–211.

Granoszewski, W., et al. 2004. "Vegetation and Climate Variability During the Last Interglacial Evidenced in the Pollen Record from Lake Baikal." *Global and Planetary Change* 46: 187–198.

Hearty, P. J., J. T. Hollin, A. C. Neumann, M. J. O'Leary, and M. Mcculloch. 2007. "Global Sea-Level Fluctuations During the Last Interglaciation (MIS 5e)." *Quaternary Science Reviews* 26: 2090–2112.

Jouzel, J., et al. 2007. "Orbital and Millennial Antarctic Climate Variability over the Past 800,000 Years." *Science* 317: 793–797.

Kaspar, F., N. Kühl, U. Cubasch, and T. Litt. 2005. "A Model-Data-Comparison of European Temperatures in the Eemian Interglacial." *Geophysical Research Letters* 32: L11703, doi:10.1029/2005GL022456.

Kühl, N., C. Gebhardt, F. Kaspar, A. Hense, and T. Litt. 2008. "Reconstruction of Quaternary Temperature Fields and Model-Data Comparison." *PAGES News* 16: 8–9.

Lozhkin, A. V., and P. M. Anderson. 1995. "The Last Interglaciation in Northeast Siberia." *Quaternary Research* 43: 147–158.

Magee, J. W., G. H. Miller, N. A. Spooner, and D. Questiaux. 2004. "A Continuous 150,000 Year Monsoon Record from Lake Eyre, Australia: Insolation Forcing Implications and Unexpected Holocene Failure." *Geology* 32: 885–888.

Marra, M. 2002. "Last Interglacial Beetle Fauna from New Zealand." *Quaternary Research* 59: 122–131.

Matthews, J. V. 1970. "Quaternary Environmental History of Interior Alaska: Pollen Samples from Organic Colluvium and Peats." *Arctic and Alpine Research* 2: 241–251.

Muller, E. H., L. Sirkin, and J. L. Craft. 1993. "Stratigraphic Evidence of a Pre-Wisconsinan Interglaciation in the Adirondack Mountains, New York." *Quaternary Research* 40: 163–168.

Müller, U. C., and G. J. Kukla. 2004. "European Environmental During the Declining Stage of the Last Interglacial." *Geology* 32: 1009–1012.

Nørgaard-Pedersen, N., N. Mikkelsen, and Y. Kristoffersen. 2009. "The Last Interglacial Warm Period Record of the Arctic Ocean; Proxy-Data Support a Major Reduction of Sea Ice." *IOP Conference Series: Earth and Environmental Science* 6: doi:10.1088/1755-1307/6/7/072002.

Péwé, T. L., G. W. Berger, J. A. Westgate, P. M. Brown, and S. W. Leavitt. 1997. "Eva Interglaciation Forest Bed, Unglaciated East-Central Alaska: Global Warming 125,000 Years Ago." Geological Society of America Special Paper 319.

Rohling, E., K. Grant, C. Hemleben, M. Siddall, B. A. A. Hoogakker, M. Bolshaw, and M. Kucera. 2009. "High Rates of Sea-Level Rise During the Last Interglacial Period." Nature Geoscience 1: 38–42.

Rundgren, M., and O. Bennike. 2002. "Century-Scale Changes of Atmospheric CO_2 During the Last Interglacial." Geology 30: 187–189.

Schweger, C. E., and , J. V. Matthews Jr. 1991. "The Last (Koy-Yukon) Interglaciation in the Yukon: Comparisons with Holocene and Interstadial Pollen Records." Quaternary International 10–12: 85–94.

Speelers, B. 2000. "The Relevance of the Eemian for the Study of the Palaeolithic Occupation of Europe." Netherlands Journal of Science 79: 283–291.

Steig, E. J., D. P. Schneider, S. D. Rutherford, M. E. Mann, J. C. Comiso, and D. T. Shindell. 2009. "Warming of the Antarctic Ice-Sheet Surface, Since the 1957 International Geophysical Year." Nature 457: 459–462.

Stirling, C. H., T. M. Esat, K. Lambeck, and M. T. Mcculloch. 1998. "Timing and Duration of the Last Interglacial: Evidence for a Restricted Interval of Widespread Coral Reef Growth." Earth and Planetary Science Letters 160: 745–762.

Stringer, C. B., J. C. Finlayson, R. N. E. Barton, Y. Fernández-Jalvo, I. Cáceres, R. C. Sabin, E. J. Rhodes, A. P. Currant, J. Rodríguez-Vidal, F. Giles-Pacheco, and J. A. Riquelme-Cantal. 2008. "Neanderthal Exploitation of Marine Mammals in Gibraltar." Proceedings of the National Academy of Sciences 105: 14319–14324.

United States Geological Survey. 2006. "Vegetation and Paleoclimate of the Last Interglacial Period, Central Alaska." http://esp.cr.usgs.gov/info/lite/alaska/alaska.html.

Vaks, A., M. Bar-Matthews, A. Ayalon, A. Matthews, L. Halicz, and A. Frumkin. 2007. "Desert Speleothems Reveal Climatic Window for African Exodus of Early Modern Humans." Geology 35: 831–834.

Van der Hammen, T., and H. Hooghiemstra. 2003. "Interglacial–Glacial Fuquene-3 Pollen Record from Colombia: An Eemian to Holocene Climate Record." Global and Planetary Change 36: 181–199.

Van Kolfschoten, T. 1992. "Aspects of the Migration of Mammals to Northwestern Europe During the Pleistocene, in Particular the Reimmigration of Arvicola Terrestris." Courier Forsch.-Inst. Senckenberg 153: 213–220.

———. 2000. "The Eemian Mammal Fauna of Central Europe." Netherlands Journal of Geosciences 79: 269–281.

Velichko, A. A., O. K. Borisova, and E. M. Zelikson. 2007. "Paradoxes of the Last Interglacial Climate: Reconstruction of the Northern Eurasia Climate Based on Palaeofloristic Data." *Boreas*, doi:10.1111/J.1502-3885.2007 .00001.X.

Walter, R. C., R. T. Buffler, J. H. Bruggeman, M. M. M. Guillaume, S. M. Berhe, B. Negassi, Y. Libsekal, H. Cheng, R. L. Edwards, R. Von Cosel, D. Néraudeau, and M. Gagnon. 2000. "Early Human Occupation of the Red Sea Coast of Eritrea During the Last Interglacial." *Nature* 405: 65–69.

Willerslev, E., et al. 2007. "Ancient Biomolecules from Deep Ice Cores Reveal a Forested Southern Greenland." *Science* 317: 111–114.

Williams, J. W., B. N. Shuman, T. Webb III, P. J. Bartlein, and P. L. Leduc. 2004. "Late-Quaternary Vegetation Dynamics in North America: Scaling from Taxa to Biomes." *Ecological Monographs* 74: 309–334.

Wilson, C. R. 2009. "A Lacustrine Sediment Record of the Last Three Interglacial Periods. From Clyde Foreland, Baffin Island, Nunavut: Biological Indicators from the Past 200,000 Years." Master's Thesis, Biology Department, Queen's University, Kingston, Ontario.

Winter, A., A. Paul, J. Nyberg, T. Oba, J. Lundberg, D. Schrag, and B. Taggart. 2003. "Orbital Control of Low-Latitude Seasonality During the Eemian." *Geophysical Research Letters* 30, doi:10.1029/2002GL016275.

4. *Life in a Super-Greenhouse*

Archer, D. 2007. "Methane Hydrate Stability and Anthropogenic Climate Change." *Biogeosciences* 4: 521–544.

Bowen, G. J., and B. B. Bowen. 2008. "Mechanisms of PETM Global Change Constrained by a New Record from Central Utah." *Geology* 36: 379–382.

―――― et al. 2002. "Mammalian Dispersal at the Paleocene/Eocene Boundary." *Science* 295: 2062–2065.

―――― et al. 2004. "A Humid Climate State During the Palaeocene/Eocene Thermal Maximum." *Nature* 432: 495–499.

Currano, E. D., P. Wilf, C. C. Labandeira, E. C. Lovelock, and D. L. Royer. 2008. "Sharply Increased Insect Herbivory During the Paleocene-Eocene Thermal Maximum." *Proceedings of the National Academy of Sciences* 105: 1060–1964.

Eberle, J. J. 2005. "A New 'Tapir' from Ellesmere Island, Arctic Canada: Implications for Northern High Latitude Palaeobiogeography and Tapir Paleobiology." *Palaeo-3* 227: 311–322.

Gibbs, S. J., P. R. Bowen, J. A. Sessa, T. J. Bralower, and P. A. Wilson. 2006. "Nannoplankton Extinction and Origination Across the Paleocene-Eocene Thermal Maximum." *Science* 314: 1770–1773.

Gingerich, P. D. 2003. "Mammalian Responses to Climate Change at the Paleocene-Eocene Boundary: Polecat Bench Record in the Northern Bighorn Basin, Wyoming." *Geological Society of America Special Paper* 369.

———. 2006. "Environment and Evolution Through the Paleocene-Eocene Thermal Maximum." *Trends in Ecology and Evolution* 21: 246–253.

Huber, M. 2008. "A Hotter Greenhouse?" *Science* 321: 353–354.

Katz, M. E., D. K. Pak, G. R. Dickens, and K. G. Miller. 1999. "The Source and Fate of Massive Carbon Input During the Latest Paleocene Thermal Maximum." *Science* 286: 1531–1533.

Kennett, J. P., and L. D. Stott. 1991. "Abrupt Deep Sea Warming, Paleoceanographic Changes and Benthic Extinctions at the End of the Paleocene." *Nature* 353: 319–322.

——— and L. D. Stott. "Global Warming." In *Effects of Past Global Change on Life.* Washington, D.C.: National Academy Press, 1995.

———, K. G. Cannariato, I. L. Hendy, and R. J. Behl. 2003. "Methane Hydrates in Quaternary Climate Change: The Clathrate Gun Hypothesis." *American Geophysical Union Special Publication* 54.

Norris, R. D., and U. Röhl. 1999. "Carbon Cycling and Chronology of Climate Warming During the Palaeocene/Eocene Transition." *Nature* 401: 775–778.

Nunes, F., and R. D. Norris. 2006. "Abrupt Reversal in Ocean Overturning During the Paleocene/Eocene Warm Period." *Nature* 439: 60–63.

Pagani, M., et al. 2006. "Arctic Hydrology During Global Warming at the Palaeocene/Eocene Thermal Maximum." *Nature* 442: 671–674.

Pearson, P. N., and M. R. Palmer. 2000. "Atmospheric Carbon Dioxide Concentrations over the Past 60 Million Years." *Nature* 406: 695–699.

——— et al. 2001. "Warm Tropical Surface Temperatures in the Late Cretaceous and Eocene Epochs." *Nature* 413: 481–487.

Royer, D. L. 2008. "Nutrient Turnover Rates in Ancient Terrestrial Ecosystems." *Palaios* 23: 421–423.

———, R. M. Kooyman, S. A. Little, and P. Wilf. 2009. "Ecology of Leaf Teeth: A Multi-Site Analysis from an Australian Subtropical Rainforest." *American Journal of Botany* 96: 738–750.

Scheibner, C., and R. P. Speijer. 2007. "Decline of Coral Reefs During Late Paleocene to Early Eocene Warming." *eEarth Discussions* 2: 133–150. http://www.electronic-earth-discuss.net/2/2007/.

Sluijs, A., et al.. 2007. "Environmental Precursors to Rapid Light Carbon Injection at the Paleocene/Eocene Boundary." *Nature* 450: 1218–1221.

Smith, T., K. D. Rose, and P. D. Gingerich. 2006. "Rapid Asia-Europe-North America Dispersal of the Earliest Eocene Primate *Teilhardina*." *Proceedings of the National Academy of Sciences USA*, 103: 11223–11227.

Sowers, T. 2006. "Late Quaternary Atmospheric CH4 Isotope Record Suggests Marine Clathrates Are Stable." *Science* 311: 838–840.

Storey, M., R. A. Duncan, and C. C. Swisher III. 2007. "Paleocene-Eocene Thermal Maximum and the Opening of the Northeast Atlantic." *Science* 316: 587–589.

Svenson, H., S. Planke, A. Malthe-Sørenssen, B. Jamtveit, R. Myklebust, T. R. Eidem, and S. S. Rey. 2004. "Release of Methane from a Volcanic Basin as a Mechanism for Initial Eocene Warming." *Nature* 429: 542–545.

Williams, C. J. 2009. "Structure, Biomass, and Productivity of a Late Paleocene Arctic Forest." *Proceedings of the Academy of Natural Sciences of Philadelphia* 158: 107–127.

—— et al. 2008. "Paleoenvironmental Reconstruction of a Middle Miocene Forest from the Western Canadian Arctic." *Palaeogrography, Palaeoclimatology, Palaeoecology* 261: 160–176.

Wing, S. L., et al. 2005. "Transient Floral Change and Rapid Global Warming at the Paleocene-Eocene Boundary." *Science* 310: 993–996.

—— et al. 2009. "Late Paleocene Fossils from the Cerrejón Formation, Colombia, Are the Earliest Record of Neotropical Rainforest." *Proceedings of the National Academy of Sciences* 106: 18627–18632.

Zachos, J. C., G. R. Dickens, and R. E. Zeebe. 2008. "An Early Cenozoic Perspective on Greenhouse Warming and Carbon-Cycle Dynamics." *Nature* 451: 279–283.

—— et al. 2001. "Trends, Rhythms and Aberrations in Global Climate 65 Ma to Present." *Science* 292: 686–693.

—— et al. 2003. "A Transient Rise in Tropical Sea Surface Temperatures During the Paleocene-Eocene Thermal Maximum." *Science* 302: 1551–1554.

—— et al. 2005. "Rapid Acidification of the Ocean During the Paleocene-Eocene Thermal Maximum." *Science* 308: 1611–1615.

5. *Future Fossils*

Bada, J. L., R. O. Peterson, A. Schimmelmann, and R. E. M. Hedges. 1990. "Moose Teeth as Monitors of Environmental Isotopic Parameters." *Oecologia* 82: 102–106.

Grimm, D. 2008. "The Mushroom Cloud's Silver Lining." *Science* 321: 1434–1437.

Kehrwald, N. M., et al.. 2008. "Mass Loss on Himalayan Glacier Endangers Water Resources." *Geophysical Research Letters* 35: L22503, doi:10.1029/2008GL035556.

Meyers, P. A. 2006. "An Overview of Sediment Organic Matter Records of Human Eutrophication in the Laurentian Great Lakes Region." *Water, Air, and Soil Pollution* 6: 89–99.

O'Reilly, C. M., S. R. Alin, P. D. Plisnier, A. S. Cohen, and B. A. Mckee. 2003. "Climate Change Decreases Aquatic Ecosystem Productivity of Lake Tanganyika, Africa." *Nature* 424: 766–768.

Ostrom, P. H., N. E. Ostrom, J. Henry, B. J. Eadie, P. A. Meyers, and J. A. Robbins. 1998. "Changes in the Trophic State of Lake Erie: Discordance Between Molecular Delta-^{13}C and Bulk Delta-^{13}C Sedimentary Records." *Chemical Geology* 152: 163–179.

Schelske, C. L., and D. A. Hodell. 1995. "Using Carbon Isotopes of Bulk Sedimentary Organic Matter to Reconstruct the History of Nutrient Loading and Eutrophication in Lake Erie." *Limnology and Oceanography* 40: 918–929.

Schmittner, A., et al. 2008. "Future Changes in Climate, Ocean Circulation, Ecosystems, and Biogeochemical Cycling Simulated for a Business-As-Usual CO_2 Emission Scenario Until Year 4000 AD." *Global Biogeochemical Cycles* 22: GB1013, doi:10.1029/2007GB002953.

Spaulding, K. L., B. A. Buchholz, L.-E. Bergman, H. Druid, and J. Frisen. 2005. "Forensics: Age Written in Teeth by Nuclear Tests." *Nature* 437: 333–334.

Totter, J. R., M. R. Zelle, and H. Hollister. 1958. "Hazard to Man of Carbon-14." *Science* 128: 1490–1495.

Verburg, P. 2007. "The Need to Correct for the Suess Effect in the Application of Delta-C13 in Sediment of Autotrophic Lake Tanganyika, as a Productivity Proxy in the Anthropocene." *Journal of Paleolimnology* 37: 591–602.

———, R. E. Hecky, and H. Kling. 2003. "Ecological Consequences of a Century of Warming in Lake Tanganyika." *Science Express* 301 (June 26): 505–507.

White, T. H. 1986. *The Once and Future King.* New York: Berkley Books.

Williams, C. P. 2007. "Recycling Greenhouse Gas Fossil Fuel Emissions into Low Radiocarbon Food Products to Reduce Human Genetic Damage." *Environmental Chemistry Letters* 5: 197–202.

6. Oceans of Acid

Anderson, N., and A. Malhoff. 1977. *The Fate of Fossil Fuel CO_2 in the Oceans.* New York: Plenum Press.

Archer, D. 2005. "Fate of Fossil Fuel CO_2 in Geologic Time." *Journal of Geophysical Research* 110: C09S05, doi:10.1029/2004JC002625.

Caldeira, K., and M. E. Wickett. 2003. "Anthropogenic Carbon and Ocean Ph." *Nature* 425: 365.

——. 2005. "Ocean Model Predictions of Chemistry Changes from Carbon Dioxide Emissions to the Atmosphere and Ocean." *Journal of Geophysical Research* 110: doi:10.1029/2004JC002671.

Cicerone, R. 2004. "The Ocean in a High CO_2 World." *Eos* 85: 351, 353.

Feeley, R. A., C. L. Sabine, J. M. Hernandez-Ayon, D. Ianson, and B. Hales. 2008. "Evidence for Upwelling of Corrosive 'Acidified' Water onto the Continental Shelf." *Science* 320: 1490–1492.

Findlay, H. S., H. L. Wood, M. A. Kendall, J. I. Spicer, R. J. Twitchett, and S. Widdicombe. 2009. "Calcification, a Physiological Process to Be Considered in the Context of the Whole Organism." *Biogeosciences Discuss* 6: 2267–2284.

Fine, M., and D. Tchernov. 2007. "Scleractinian Coral Species Survive and Recover from Decalcification." *Science* 315: 1811.

Gibbs, S. J., P. R. Bowen, J. A. Sessa, T. J. Bralower, and P. A. Wilson. 2006. "Nannoplankton Extinction and Origination Across the Paleocene-Eocene Thermal Maximum." *Science* 314: 1770–1773.

Guinotte, J. M., J. Orr, S. Cairns, A. Friewald, L. Morgan, and R. George. 2006. "Will Human-Induced Changes in Seawater Chemistry Alter the Distribution of Deep-Sea Scleractinian Corals?" *Frontiers in Ecology and the Environment* 4: 141–146.

Hall-Spencer, J. M., R. Rodolfo-Metalpa, S. Martin, E. Ransome, M. Fine, S. M. Turner, S. J. Rowley, D. Tedesco, and M.-C. Buia. 2008. "Volcanic Carbon Dioxide Vents Show Ecosystem Effects of Ocean Acidification." *Nature* 454: 96–99.

Henderson, C. 2006. "Paradise Lost. *New Scientist* 5: 29–33.

Hoegh-Guldberg, O., et al. 2007. "Coral Reefs Under Rapid Climate Change and Ocean Acidification." *Science* 318: 1737–1742.

Iglesias-Rodriguez, M. D., P. R. Halloran, R. E. M. Rickaby, I. R. Hall, E. Colmenero-Hidalgo, J. R. Gittins, D. R. H. Green, T. Tyrrell, S. J. Gibbs, P. Von Dassow, E. Rehm, E. V. Armbrust, K. P. Boessenkool. 2008. "Phytoplankton Calcification in a High-CO_2 World." *Science* 320: 336–340.

Interacademy Panel on International Issues. 2009. "IAP Statement on Ocean Acidification," June 2009. http://www.interacademies.net/object.file/master/9/075/statement_RS1579_IAP_05.09final2.pdf.

Kleypas, J. A., R. W. Buddemeier, D. Archer, J.-P. Gattuso, C. Langdon, and B. N. Opdyke. 1999. "Geochemical Consequences of Increased Atmospheric Carbon Dioxide on Coral Reefs." *Science* 284: 118–120.

Kolbert, E. 2006. "The Darkening Sea." *The New Yorker,* November 20.

Morel, V. 2007. "Into the Deep: First Glimpses of Bering Sea Canyons Heats Up Fisheries Battle." *Science* 318: 181–182.

Orr, J. C., et al. 2005. "Anthropogenic Ocean Acidification over the Twenty-First Century and Its Impact on Calcifying Organisms." *Nature* 437: 681–686.

Poore G. C. B., and G. Wilson. 1993. "Marine Species Richness." *Nature* 361: 597–598.

Precht, W. F., and R. B. Aronson. 2004. "Climate Flickers and Range Shifts of Reef Corals." *Frontiers in Ecology and the Environment* 2: 307–314.

Richardson, A. J., and M. J. Gibbons. 2008. "Are Jellyfish Increasing in Response to Ocean Acidification?" *Limnology and Oceanography* 53: 2040–2045.

Rintoul, S. R. 2007. "Rapid Freshening of Antarctic Bottom Water in the Indian and Pacific Oceans." *Geophysical Research Letters* 34: L06606. 1-L06606. 5.

Roberts, J. M., A. J. Wheeler, and A. Friewald. 2006. "Reefs of the Deep: The Biology and Geology of Cold-Water Coral Ecosystems." *Science* 312: 543–547.

Roberts, S., and M. Hirschfield. 2004. "Deep-Sea Corals: Out of Sight, But Not Out of Mind." *Frontiers in Ecology and the Environment* 2: 123–130.

Royal Society. 2005. "Ocean Acidification Due to Increasing Atmospheric Carbon Dioxide." Royal Society Policy Document 12/05. http://www.us-ocb.org/publications/Royal_Soc_OA.pdf.

Sabine, C. L. 2004. "The Oceanic Sink for Anthropogenic CO_2." *Science* 305: 367–371.

Scheibner, C., and R. P. Speijer. 2007. "Decline of Coral Reefs During Late Paleocene to Early Eocene Warming." *eEarth Discussions* 2: 133–150. http://www.electronic-earth-discuss.net/2/2007/.

Silverman, J., B. Lazar, L. Cao, K. Caldeira, and J. Erez. 2009. "Coral Reefs May Start Dissolving When Atmospheric CO_2 Doubles." *Geophysical Research Letters* 36: L05606, doi:10.1029/2008GL036282.

Steinacher, M., F. Joos, T. L. Frolicher, G.-K. Plattner, and S. C. Doney. 2009. "Imminent Ocean Acidification in the Arctic Projected with the NCAR Global Coupled Carbon Cycle-Climate Model." *Biogeosciences* 6: 515–533.

UNEP. 2006. "Marine and Coastal Ecosystems and Human Well-Being: A Synthesis Report Based on the Findings of the Millennium Ecosystem Assessment."

Yamamoto-Kawai, M., F. A. Mclaughlin, E. C. Carmack, S. Nishino, and K. Shimada. 2009. "Aragonite Undersaturation in the Arctic Ocean: Effects of Ocean Acidification and Sea Ice Melt." *Science* 326: 1098–1100.

Zachos, J. C., U. Röhl, S. A. Schellenberg, A. Sluijs, D. A. Hodell, D. C. Kelly, E. Thomas, M. Nicolo, I. Raffi, L. J. Lourens, H. Mccarren, and D. Kroon. 2005. "Rapid Acidification of the Ocean During the Paleocene-Eocene Thermal Maximum." *Science* 308: 1611–1615.

7. The Rising Tide

Alexander, C. 2008. "Tigerland." *The New Yorker*, April 21: 67–73.

Alley, R. B., P. U. Clark, P. Huybrechts, and I. Joughin. 2005. "Ice-Sheet and Sea-Level Changes." *Science* 310: 456–460.

Ballard, R. D., D. F. Coleman, and G. D. Rosenberg. 2000. "Further Evidence of Abrupt Holocene Drowning of the Black Sea Shelf." *Marine Geology* 170: 253–261.

Bamber, J. L., R. E. M. Riva, B. L. A. Vermeersen, and A. M. Lebrocq. 2009. "Reassessment of the Potential Sea-Level Rise from a Collapse of the West Antarctic Ice Sheet." *Science* 324: 901–903.

Bentley, C. R. 1997. "Rapid Sea-Level Rise Soon from West Antarctic Ice Sheet Collapse?" *Science* 275: 1077–1078.

Blanchon, P., A. Eisenhauer, J. Fietzke, and V. Liebetrau. 2009. "Rapid Sea-Level Rise and Reef Back-Stepping at the Close of the Last Interglacial Highstand." *Nature* 458: 881–884.

Bo, S., M. J. Siegert, S. M. Mudd, D. Sugden, S. Fujita, C. Xiangbin, J. Yunyun, T. Xueyuan, and L. Yuansheng. 2009. "The Gamburtsev Mountains and the Origin and Early Evolution of the Antarctic Ice Sheet." *Nature* 459: 690–693.

Böcker, A. 1998. *Regulation of Migration: International Experiences.* Antwerp, Belgium: Het Spinhuis.

Cabanes, C., A. Cazenave, and C. Le Provost. 2001. "Sea Level Rise During Past 40 Years Determined from Satellite and In Situ Observations." *Science* 294: 840–842.

Cazenave, A. 2006. "How Fast Are the Ice Sheets Melting?" *Science* 314: 1250–1252.

Cazenave, A., K. Dominh, S. Guinehut, E. Berthier, W. Llovel, G. Ramillien, M. Ablain, and G. Larnicol. 2008. "Sea Level Budget over 2003–2008: A Reevaluation from GRACE Space Gravimetry, Satellite Altimetry and Argo." *Global and Planetary Change,* doi:10.1016/J/.Gloplacha .2008.10.004.

Church, J. A., J. S. Godfrey, D. R. Jackett, and T. Mcdougall. 1991. "A Model of Sea Level Rise Caused by Ocean Thermal Expansion." *Journal of Climate* 4: 438–456.

Clark, P. U., A. M. Mccabe, A. C. Mix, and A. J. Weaver. 2004. "Rapid Rise of Sea Level 19,000 Years Ago and Its Global Implications." *Science* 304: 1141–1144.

Davis, C. H., Y. Li, J. R. Mcconnell, M. M. Frey, and E. Hanna. 2005. "Snowfall-Driven Growth in East Antarctic Ice Sheet Mitigates Recent Sea-Level Rise." *Science* 308: 1898–1901.

Garcia-Castellano, D., F. Estrada, I. Jiménez-Munt, C. Gorini, M. Fernàndez, J. Vergés, and R. De Vicente. 2009. "Catastrophic Flood of the Mediterranean After the Messinian Salinity Crisis." *Nature* 462: 778–781.

Hansen, J. E. 2007. "Scientific Reticence and Sea Level Rise." *Environmental Research Letters* 2, doi:10.1088/1748-9326/2/2/024002.

Hu, A., G. A. Meehl, W. Han, and J. Yin. 2009. "Transient Response of the MOC and Climate to Potential Melting of the Greenland Ice Sheet in the 21st Century." *Geophysical Research Letters* 36: L10707, doi:10.1029/2009GL037998.

Joughin, I., S. B. Das, M. A. King, B. E. Smith, I. M. Howatt, and T. Moon. 2008. "Seasonal Speedup Along the Western Flank of the Greenland Ice Sheet." *Science* 320: 781–783.

Karan, P. P. 2005. *Japan in the 21st Century: Environment, Economy, and Society.* Lexington: University Press of Kentucky.

Kellogg, W. W., and M. Mead, eds. 1976. *The Atmosphere: Endangered and Endangering.* Fogarty International Center Proceedings, no. 39. DHEW Publication Number (NIH) 77–1065.

Kerr, R. 2007. "How Urgent Is Climate Change?" *Science* 318: 1230–1231.

Khan, S. A., P. Knudsen, and C. C. Tscherning. 2003. "Crustal Deformations at Permanent GPS Sites in Denmark." In: *Window on the Future of Geodesy*, F. Sanso, ed. Berlin: Springer. doi:10.1007/3-540-27432-4_94.

Larter, R. D., et al. 2007. "West Antarctic Ice Sheet Change Since the Last Glacial Period." *Eos* 88: 189–190.

Liu, G., Luo, X., Q. Chen, D. Huang, and X. Ding. 2008. "Detecting Land Subsidence in Shanghai by PS-Networking SAR Interferometry." *Sensors* 8: 4725–4741.

Marbaix, P., and R. J. Nichols. 2007. "Accurately Determining the Risks of Rising Sea Level." *Eos* 88: 441–442.

Meehl, G. A., et al. 2005. "How Much More Global Warming and Sea Level Rise?" *Science* 307: 1769–1772.

Mitrovica, J. X., N. Gomez, and P. U. Clark. 2009. "The Sea-Level Fingerprint of West Antarctic Collapse." *Science* 323: 753.

Oerlemans, J., D. Dahl-Jensen, and V. Masson-Delmotte. 2006. "Ice Sheets and Sea Level. Letter to *Science*." *Science* 313: 1043–1044.

Pfeffer, W. T., J. T. Harper, and S. O'Neel. 2008. "Kinematic Constraints on Glacier Contributions to 21st-Century Sea-Level Rise." *Science* 321: 1340–1343.

Pollard, D., and R. M. Deconto. 2009. "Modelling West Antarctic Ice Sheet Growth and Collapse Through the Past Five Million Years." *Nature* 458: 329–332.

Pritchard, H. D., R. J. Arthern, D. G. Vaughan, and L. A. Edwards. 2009. "Extensive Dynamic Thinning on the Margins of the Greenland and Antarctic Ice Sheets." *Nature* 461: 971–975.

Rohling, E. J., K. Grant, C. Hemleben, M. Siddall, B. A. A. Hoogakker, M. Bolshaw and M. Kucera. 2008. "High Rates of Sea-Level Rise During the Last Interglacial Period." *Nature Geoscience* 1: 38–42.

Rowley, J., et al. 2007. "Risk of Rising Sea Level to Population and Land Area." *Eos* 88, 105.

Ryan, W. B. F., et al. 1997. "An Abrupt Drowning of the Black Sea Shelf." *Marine Geology* 138, 119–126.

Shepherd, A., and D. Wingham. 2007. "Recent Sea-Level Contributions of the Antarctic and Greenland Ice Sheets." *Science* 315: 1529–1532.

Siddall, M. 2009. "The Sea Level Conundrum: Insights from Paleo Studies." *Eos* 90: 72–73.

Steig, E. J., and A. P. Wolfe. 2008. "Sprucing Up Greenland." *Science* 320: 1595–1596.

———, D. P. Schneider, S. D. Rutherford, M. E. Mann, J. C. Comiso, and D. T. Shindell. 2009. "Warming of the Antarctic Ice-Sheet Surface Since the 1957 International Geophysical Year." *Nature* 457: 459–462.

Stockstad, E. 2007. "Boom and Bust in a Polar Hot Zone." *Science* 315: 1522–1523.

Velicogna, I., and J. Wahr. 2006. "Measurements of Time-Variable Gravity Show Mass Loss in Antarctica." *Science* 311: 1754–1756.

Wigley, T. M. L. 2005. "The Climate Change Commitment." *Science* 307: 1766–1769.

Xuo, Y.-Q., Y. Zhang, S.-J. Ye, J.-C. Wu, and Q.-F. Li. 2005. "Land Subsidence in China." *Environmental Geology* 48: 713–720.

Yanko-Hombach, V. 2003. "'Noah's Flood' and the Late Quaternary History of the Black Sea and Its Adjacent Basins: A Critical Overview of the Flood Hypotheses." Geological Society of America Annual Meeting, Seattle. *Abstracts with Programs* 36: 460.

Yu, S.-Y., Y.-X. Li, and T. E. Törnqvist. 2009. "Tempo of Global Deglaciation During the Early Holocene: A Sea Level Perspective." *PAGES News* 17: 68–70.

8. An Ice-Free Arctic

ACIA. 2005. "Arctic Climate Impact Assessment." Cambridge, UK: Cambridge University Press.

Borgerson, S. G. 2008. "Arctic Meltdown." *Foreign Affairs* (March/April): 63–77.

Briner, J. P., N. Michelutti, D. R. Francis, G. H. Miller, Y. Axford, M. J. Wooller, and A. P. Wolfe. 2006. "A Multi-Proxy Lacustrine Record of Holocene Climate Change on Northeastern Baffin Island, Arctic Canada." *Quaternary Research* 65: 431–442.

Carlton, J. 2005. "Is Global Warming Killing the Polar Bears?" *Wall Street Journal*, December 14, 2005.

CBC News. 2006. "Scientists Fear Disease Outbreaks in Northern Whales." http://Cbc.Ca/Canada/North/Story/2006/08/16/Whale-Disease-North.html.

Census of Marine Life, Arctic Ocean Diversity, School of Fisheries and Ocean Sciences, University of Alaska, Fairbanks. http://www.arcodiv.org/index.html.

Chylek, P., M. K. Dubey, and G. Lesins. 2006. "Greenland Warming of 1920–1930 and 1995–2005." *Geophysical Research Letters* 33: L11707, doi:10.1029/2006GL026510.

Cressey, D. 2008. "The Next Land Rush." *Nature* 451: 12–15.

Derocher, A. E., et al. 2002. "Diet Composition of Polar Bears in Svalbard and the Western Barents Sea." *Polar Biology* 25: 448–452.

Derocher, A. E., et al. 2000. "Predation of Svalbard Reindeer by Polar Bears." *Polar Biology* 23: 675–678.

Douglas, M. S. V., J. P. Smol, and W. Blake, Jr. 1994. "Marked Post-18th Century Change in High-Arctic Ecosystems." *Science* 266: 416–419.

Fisher, D., A. Dyke, R. Koerner, J. Bourgeois, C. Kinnard, C. Zdanowicz, A. De Vernal, C. Hillaire-Marcel, J. Savelle, and A. Rochon. 2006. "Natural Variability of Arctic Sea Ice over the Holocene," *Eos Trans. AGU*, 87: 273.

Gaston, A. J., and K. Woo. 2008. "Razorbills (*Alca Torda*) Follow Subarctic Prey into the Canadian Arctic: Colonization Results from Climate Change?" *Auk* 125: 939–942.

Gautier, D. L., et al.. 2009. "Assessment of Undiscovered Oil and Gas in the Arctic." *Science* 324: 1175–1179.

Grahl-Nielsen, O., et al. 2003. "Fatty Acid Composition of the Adipose Tissue of Polar Bears and of Their Prey: Ringed Seals, Bearded Seals, and Harp Seals." *Marine Ecology Progress Series* 265: 275–282.

Grebmeier, J. M., et al. 2006. "A Major Ecosystem Shift in the Northern Bering Sea." *Science* 311: 1461–1464.

Iredale, W. 2005. "Polar Bears Drown as Ice Shelf Melts." *Sunday Times* [London], December 18.

International Arctic Science Committee. 2008. "Latitudinal Gradients in Species Diversity in the Arctic." In: *Encyclopedia of Earth,* M. Mcginley, topic ed. http://www.eoearth.org/article/latitudinal_gradients_in_species_ diversity_in_the_arctic.

Kaufman, D. S., et al. 2004. "Holocene Thermal Maximum in the Western Arctic (0–180° W)." *Quaternary Science Reviews* 23: 529–560.

Lomborg, B. 2007. *Cool It.* New York: Knopf.

Lowenstein, T. K., and R. V. Demicco. 2006. "Elevated Eocene Atmospheric CO and Its Subsequent Decline." *Science* 313: 1928.

Mellgren, D. 2007. "Technology, Climate Change Spark Race to Claim Arctic Resources." *USA Today.* http:www.usatoday.com/money/world/2007-03 -24-arcticbonanza_n.htm.

Monnett, C., and J. S. Gleason. 2006. "Observations of Mortality Associated with Extended Open-Water Swimming by Polar Bears in the Alaskan Beaufort Sea." *Polar Biology* 29: 681–687.

Moore, P. D. 2004. "Hope in the Hills for Tundra?" *Nature* 432: 159.

Mieszkowska, N., D. Sims, and S. Hawkins. 2007. "Fishing, Climate Change and North-East Atlantic Cod Stocks." Report for the World Wildlife Fund, UK.

Overland, J., J. Turner, J. Francis, N. Gillett, G. Marshall, and M. Tjernström. 2008. "The Arctic and Antarctic: Two Faces of Climate Change." *Eos* 89: 177.

Putkonen, J., T. C. Grenfell, K. Rennert, C. Bitz, P. Jacobson, and D. Russell. 2009. "Rain on Snow: Little Understood Killer in the North." *Eos* 26: 221–222.

Rigor, I. G., and J. M. Wallace. 2004. "Variations in the Age of Arctic Sea-Ice and Summer Sea-Ice Extent." *Geophysical Research Letters* 31: L09401, doi:10. 1029/2004GL019492.

Schiermeier, Q. 2007. "The New Face of the Arctic." *Nature* 446: 133–135.

Serreze, M. C., M. M. Holland, and J. Stroeve. 2007. "Perspectives on the Arctic's Shrinking Sea-Ice Cover." *Science* 315: 1533–1536.

Smol, J. P., and M. S. V. Douglas. 2007. "Crossing the Final Ecological Threshold in High Arctic Ponds." *Proceedings of the National Academy of Sciences* 104: 12395–12397.

Stickley, C., K. St. John, N. Koç, R. W. Jordan, S. Passchier, R. B. Pearce, and L. E. Kearns. 2009. "Evidence for Middle Eocene Arctic Sea Ice from Diatoms and Ice-Rafted Debris." *Nature* 460: 376–379.

Stirling, I., and A. E. Derocher. 2007. "Melting Under Pressure: The Real Scoop on Climate Warming and Polar Bears." *Wildlife Professional*, Fall.

Stirling, I., and A. E. Derocher. 1993. "Possible Impacts of Climatic Warming on Polar Bears." *Arctic* 46: 240–245.

Stockstad, E. 2007. "Boom and Bust in a Polar Hot Zone." *Science* 315: 1522–1523.

Stroeve, J. M., et al. 2008. "Arctic Sea Ice Extent Plummets in 2007." *Eos* 89: 13.

Torrice, M. 2009. "Science Lags on Saving the Arctic from Oil Spills." *Science* 325: 1335.

Tripati, A. K., C. D. Roberts, and R. A. Eagle. 2009. "Coupling of CO_2 and Ice Sheet Stability over Major Climate Transitions of the Last 20 Million Years." *Science* 326: 1394–1397.

Wilkinson, J. P., P. Gudmandsen, S. Hanson, R. Saldo, and R. M. Samuelson. 2009. "Hans Island: Meteorological Data from an International Border-line." *Eos* 90: 190–191.

Woods Hole Oceanographic Institution. 2006. "Walrus Calves Stranded by Melting Sea Ice." *Science Daily,* April 15.

World Climate Report. 2007 "More on Polar Bears." http://www.worldclimate report.com/index.php/2007/12/12/more-on-polar-bears/.

9. The Greening of Greenland

Alley, R. B., P. U. Clark, P. Huybrechts, and I. Joughin. 2005. "Ice Sheet and Sea Level Changes." *Science* 310: 456–460.

Bamber, J. L., R. L. Layberry, and S. P. Gogenini. 2001. "A New Ice Thickness and Bed Data Set for the Greenland Ice Sheet 1: Measurement, Data Reduction, and Errors." *Journal of Geophysical Research* 106 (D24): 33773–33780.

————. 2001. "A New Ice Thickness and Bed Data Set for the Greenland Ice Sheet 2: Relationship Between Dynamics and Basal Topography." *Journal of Geophysical Research* 106 (D24): 33781–33788.

Blanchon, P., A. Eisenhauer, J. Fietzke, and V. Liebetrau. 2009. "Rapid Sea-Level Rise and Reef Back-Stepping at the Close of the Last Interglacial High-stand." *Nature Geoscience* 458: 881–885.

Cazenave, A. 2006. "How Fast Are the Ice Sheets Melting?" *Science* 314: 1250–1252.

Charpentier, R. R., T. R. Klett, and E. D. Attanasi. 2008. "Database for Assessment Unit-Scale Analogs (Exclusive of the United States): U. S. Geological Survey." Open-File Report 2007-1404. http://geology.com/usgs/ Arctic-Oil-And-Gas-Report.shtml.

Cuffey, K., and S. Marshall. 2000. "Substantial Contribution to Sea-Level Rise During the Last Interglacial from the Greenland Ice Sheet." *Nature* 404: 591–594.

De Vernal, A., and C. Hillaire-Marcel. 2008. "Natural Variability of Greenland Climate, Vegetation, and Ice Volume During the Past Million Years." *Science* 320: 1622–1625.

Francis, D. R., A. P. Wolfe, I. R. Walker, and G. H. Miller. 2006. "Interglacial and Holocene Temperature Reconstructions Based on Midge Remains in Sediments of Two Lakes from Baffin Island, Nunavut, Arctic Canada." *Palaeo-3* 236: 107–124.

Gregory, J. M., P. Huybrechts, and S. C. B. Raper. 2004. "Threatened Loss of the Greenland Ice-Sheet." *Nature* 428: 616.

Hamilton, G. S., V. B. Spikes, and L. A. Stearns. 2005. "Spatial Patterns in Mass Balance of the Siple Coast and Amundsen Sea Sectors, West Antarctica." *Annals of Glaciology* 41: 105–106.

Hanna, E., P. Huybrechts, I. Janssens, J. Cappelen, K. Steffen, and A. Stephens. 2005. "Runoff and Mass Balance of the Greenland Ice Sheet: 1958–2003." *Journal of Geophysical Research* 110: D13108, doi:10.1029/2004JD005641.

Huybrechts, P., and J. De Wolde. 1999. "The Dynamic Response of the Greenland and Antarctic Ice Sheets to Multiple-Century Climatic Warming." *Journal of Climate* 12: 2169–2188.

IPCC. 2007. "Summary for Policymakers." In: *Climate Change 2007: Impacts, Adaptation and Vulnerability. Contribution of Working Group II to the Fourth Assessment Report of the Intergovernmental Panel on Climate Change*, M. L. Parry et al., eds. Cambridge, UK: Cambridge University Press.

Kerr, R. A. 2008. "Winds, Not Just Global Warming, Eating Away at the Ice Sheets." *Science* 322: 33.

Khan, S. A., J. Wahr, L. A. Stearns, G. S. Hamilton, T. Van Dam, K. M. Larson, and O. Francis. 2007. "Elastic Uplift in Southeast Greenland Due to Rapid Ice Mass Loss." *Geophysical Research Letters* 34: L21701, doi:10.1029/2007GL031468.

Letreguilly, A., P. Huybrechts, and N. Reeh. 1991. "Steady-State Characteristics of the Greenland Ice Sheet Under Different Climates." *Journal of Glaciology* 37: 149–157.

Luthcke, S., H. J. Zwally, W. Abdalati, D. D. Rowlands, R. D. Ray, R. S. Nerem, F. G. Lemoine, J. J. Mccarthy, and D. S. Chinn. 2006. "Recent Greenland Ice Mass Loss by Drainage System from Satellite Gravity Observations." *Science* 314: 1286.

Oerlemans, J., D. Dahl-Jensen, and V. Masson-Delmotte. 2006. "Ice Sheets and Sea Level: Letter to *Science*." *Science* 313: 1043–1044.

Overpeck, J. T., B. L. Otto-Bliesner, G. H. Miller, D. R. Muhs, R. B. Alley, and
J. T. Kiehl. 2006. "Paleoclimatic Evidence for Future Ice-Sheet Instability
and Rapid Sea-Level Rise." *Science* 311: 1747–1750.

Pritchard, H. D., R. J. Arthern, D. G. Vaughan, and L. A. Edwards. 2009. "Extensive Dynamic Thinning on the Margins of the Greenland and Antarctic
Ice Sheets." *Nature* 461: 971–975.

Rial, J. A., C. Tang, and K. Steffen. "Glacial Rumblings from Jakobshavn Ice
Stream, Greenland." *Journal of Glaciology* 55: 389–399.

Ridley, J. K., P. Huybrechts, J. M. Gregory, and J. A. Lowe. 2005. "Elimination of
the Greenland Ice Sheet in a High CO_2 Climate." *Journal of Climate* 18:
3409–3427.

Schwartz, M. L. 2005. *Encyclopedia of Coastal Science*. Berlin: Springer.

Secher, K., and P. Appel. 2007. "Gemstones of Greenland." *Geology and Ore* 7: 1–12.

Shepherd, A., and D. Wingham. 2007. "Recent Sea-Level Contributions of the
Antarctic and Greenland Ice Sheets." *Science* 315: 1529–1532.

Stearns, L. A., and G. S. Hamilton. 2007. "Rapid Volume Loss from Two East
Greenland Outlet Glaciers Quantified Using Repeat Stereo Satellite Imagery." *Geophysical Research Letters* 34: in press, doi:10.1029/2006GL028982.

Steig, E. J., and A. P. Wolfe. 2008. "Sprucing Up Greenland." *Science* 320:
1595–1597.

Steig, E. J., D. P. Schneider, S. D. Rutherford, M. E. Mann, J. C. Comiso, and
D. T. Shindell. 2009. "Warming of the Antarctic Ice-Sheet Surface Since
the 1957 International Geophysical Year." *Nature* 457: 459–462.

Traufetter, G. 2006. "Global Warming a Boon for Greenland's Farmers." *Spiegel
International Online*, August 30.

Truffer, M., and M. Fahnestock. 2007. "Rethinking Ice Sheet Time Scales." *Science* 315: 1508–1510.

Van den Broeke, M., J. Bamber, J. Ettema, E. Rignot, E. Schrama, W. J. Van De
Berg, E. Van Meijgaard, I. Velicogna, and B. Wouters. 2009. "Partitioning
Recent Greenland Mass Loss." *Science* 326: 984–986.

Zwally, H. J., W. Abdalati, T. Herring, K. Larson, J. Saba, and K. Steffen. 2002.
"Surface Melt-Induced Acceleration of Greenland Ice-Sheet Flow." *Science* 297: 218–222.

10. *What About the Tropics?*

Allison, I., et al.. 2009. *The Copenhagen Diagnosis: Updating the World on the Latest
Climate Science*. Sydney, Australia: University of New South Wales Climate Change Research Centre.

Bar-Matthews, M., A. Ayalon, and A. Kaufman. 2000. "Timing and Hydrological Conditions of Sapropel Events in the Eastern Mediterranean, as Evident from Speleothems, Soreq Cave, Israel." *Chemical Geology* 169: 145–156.

Behling, H., G. Keim, G. Irion, W. Junk, and J. Nunes De Mello. 2001. "Holocene Environmental Changes in the Central Amazon Basin Inferred from Lago Calado (Brazil)." *Palaeogeography, Palaeoclimatology, Palaeoecology* 173: 87–101.

Biastoch, A., C. W. Boning, F. U. Schwarzkopf, and J. R. E. Letjeharms. 2009. "Increase in Agulhas Leakage Due to Poleward Shift of Southern Hemisphere Westerlies." *Nature* 462: 495–498.

Boko, M., et al. 2007. "Africa." In: *Climate Change 2007: Impacts, Adaptation and Vulnerability. Contribution of Working Group II to the Fourth Assessment Report of the Intergovernmental Panel on Climate Change*, M. L. Parry et al., eds. Cambridge, UK: Cambridge University Press.

Cherry, M. 2005. "Ministers Agree to Act on Warmings of Soaring Temperatures in Africa." *Nature* 437: 1217.

Christensen, J. H., et al. 2007. "Regional Climate Projections." In: *Climate Change 2007: The Physical Science Basis. Contribution of Working Group I to the Fourth Assessment Report of the Intergovernmental Panel on Climate Change*, S. Solomon et al., eds. Cambridge, UK: Cambridge University Press.

Collins, M., and the CMIP Modelling Groups. 2005. "El Niño: Or La Niña-Like Climate Change?" *Climate Dynamics* 24: 89–104.

Demenocal, P. B., J. Adkins, J. Ortiz, and T. Guilderson. 2000. "Millennial-Scale Sea Surface Temperature Variability During the Last Interglacial and Its Abrupt Termination." *American Geophysical Union 2000 Abstracts*.

Dessai, S., M. Hulme, R. Lempert, and R. Pielke. 2009. "Do We Need Better Predictions to Adapt to a Changing Climate?" *Eos* 90: 111–112.

Easterbrook, G. 2007. "Global Warming: Who Loses—and Who Wins." *The Atlantic* 299 (3): 52–64.

Eschenbach, W. 2004. "Climate-Change Effect on Lake Tanganyika?" *Nature* 430, doi:10.1038/Nature02689.

Funk, C., M. D. Dettinger, J. C. Michaelsen, J. P. Verdin, M. E. Brown, M. Barlow, and A. Hoell. 2008. "Warming of the Indian Ocean Threatens Eastern and Southern African Food Security But Could Be Mitigated by Agricultural Development." *Proceedings of the National Academy of Sciences* 105: 11081–11086.

Giles, J. 2007. "How to Survive a Warming World." *Nature* 446: 716–717.

Hay, S. I., J. Cox, D. J. Rogers, S. E. Randolph, D. I. Stern, G. D. Shanks, M. F. Myers, and R. W. Snow. 2002. "Climate Change and the Resurgence of Malaria in the East African Highlands." *Nature* 415: 905–909.

Huber, M., and R. Caballero. 2003. "Eocene El Niño: Evidence for Robust Tropical Dynamics in the "Hothouse." *Science* 299: 877–881.

Huey, R. B., C. A. Deutsch, J. J. Tewksbury, L. J. Vitt, P. E. Hertz, H. J. Alvarez Perez, and T. Garland, Jr. 2009. "Why Tropical Forest Lizards Are Vulnerable to Climate Warming." *Proceedings of the Royal Society*, B, 106: 3547–3648.

Hulme, M., R. Doherty, T. Ngara, M. New, and D. Lister. 2000. "African Climate Change: 1900–2100." *Climate Research* 17: 145–168.

Landsea, C. W. 2007. "Counting Atlantic Tropical Cyclones Back to 1900." *Eos* 88: 197–202.

Linthicum, K. J., A. Anyamba, C. J. Tucker, P. W. Kelley, M. F. Myers, and C. L. Peters. 1999. "Climate and Satellite Indicators to Forecast Rift Valley Fever Epidemics in Kenya." *Science* 285: 397–400.

Lobell, D. B., et al. 2008. "Prioritizing Climate Change Adaptation Needs for Food Security in 2030." *Science* 319: 607–610.

Malhi, Y., J. T. Roberts, R. A. Betts, T. J. Kileen, W. Li, and C. A. Nobre. 2008. "Climate Change, Deforestation, and the Fate of the Amazon." *Science* 319: 169–172.

Mayle, F. E., and M. J. Power 2008. "Impacts of a Drier Early-Mid-Holocene Climate Upon Amazonian Forests." *Philos Trans R Soc Lond B Biol Sci.* 363(1498): 1829–1838.

Mote, P. W., and G. Kaser. 2007. "The Shrinking Glaciers of Kilimanjaro: Can Global Warming Be Blamed?" *American Scientist* 95: 318–325.

Moy, C. M., G. O. Seltzer, D. T. Rodbell, and D. M. Anderson. 2002. "Variability of El Niño/Southern Oscillation Activity at Millennial Timescales During the Holocene Epoch." *Nature* 420, 162–165.

Neelin, J. D., M. Munnich, H. Su, J. E. Meyerson, and C. E. Holloway. 2006. "Tropical Drying Trends in Global Warming Models and Observations." *Proceedings of the National Academy of Sciences* 103: 6110–6115.

O'Reilly, C. M., S. R. Alin, P. D. Plisnier, A. S. Cohen, and B. A. Mckee. 2003. "Climate Change Decreases Aquatic Ecosystem Productivity of Lake Tanganyika, Africa." *Nature* 424: 766–768.

Patt, A. G., L. Ogallo, and M. Hellmuth. 2007. "Learning from 10 Years of Climate Outlook Forums in Africa." *Science* 318: 49.

Ramaswamy, V. 2009. "Anthropogenic Climate Change in Asia: Key Challenges." *Eos* 90: 469–471.

Russell, J., H. Eggermont, R. Taylor, and D. Verschuren. 2008. "Paleolimnological Records of Recent Glacier Recession in the Ruwenzori Mountains, Uganda-D. R. Congo." *Journal of Paleolimnology*, doi:10.1007/S10933-008-9224-4.

Sarewitz, D. 2010. "Tomorrow Never Knows." *Nature* 463: 24.

Saunders, M. A., and A. S. Lea. 2008. "Large Contribution of Sea Surface Warming to Recent Increase in Atlantic Hurricane Activity." *Nature* 451: 557–559.

Scholz, C. A., et al. 2007. "East African Megadroughts Between 135 and 75 Thousand Years Ago and Bearing on Early-Modern Human Origins." *Proceedings of the National Academy of Sciences* 104: 16416–16421.

Seidel, D. J., and W. J. Randel. 2007. "Recent Widening of the Tropical Belt: Evidence from Tropopause Observations." *Journal of Geophysical Research* 112: D20113, doi:10.1029/2007JD008861.

Shukla, J. 2007. "Monsoon Mysteries." *Science* 318: 204–205.

Stager, J. C., B. Cumming, and L. D. Meeker. 2003. "A 10,200-Year High-Resolution Diatom Record from Pilkington Bay, Lake Victoria, Uganda." *Quaternary Research* 59: 172–181.

—— et al. 2005. "Solar Variability and the Levels of Lake Victoria, East Africa, During the Last Millennium." *Journal of Paleolimnology* 33: 243–251.

—— et al. 2007. "Solar Variability, ENSO, and the Levels of Lake Victoria, East Africa." *Journal of Geophysical Research* 112: D15106, doi:10.1029/2006JD008362.

Tadross, M., C. Jack, and D. Le Sueur. 2005. "On RCM-Based Projections of Change in Southern African Summer Climate." *Geophysical Research Letters* 32: L23713, doi:10.1029/2005GL024460.

Thomas, C. 2004. "Changed Climate in Africa?" *Nature* 427: 690–691.

USAID. 2003. "Rainfall in Ethiopia Is Becoming Increasingly Erratic." http://www.reliefweb.int/rw/rwb.nsf/alldocsbyunid/66e46860182a84da49256d5100071e16.

Vallely, P. 2006. "Climate Change Will Be Catastrophe for Africa." *The Independent*, May 16.

Verburg, P., R. E. Hecky, and H. Kling. 2003. "Ecological Consequences of a Century of Warming in Lake Tanganyika." *Science Express* 301: 505–507.

Verschuren, D., K. R. Laird, and B. F. Cumming. 2000. "Rainfall and Drought in Equatorial East Africa During the Past 1,100 Years." *Nature* 403: 410–413.

—— et al. 2009. "Half-Precessional Dynamics of Monsoon Rainfall Near the East African Equator." *Nature* 462: 637–641.

Vuille, M., B. Francou, P. Wagnon, I. Juen, G. Kaser, B. G. Mark, and R. S. Bradley. 2008. "Climate Change and Tropical Andean Glaciers: Past, Present, and Future." *Earth Science Reviews* 89: 79–96.

11. Bringing It Home

Bagla, P. 2009. "No Sign Yet of Himalayan Meltdown, Indian Report Finds." *Science* 326: 924–925.

Cannone, N., S. Sgorbati, and M. Guglielmin. 2007. "Unexpected Impacts of Climate Change on Alpine Vegetation." *Frontiers in Ecology and the Environment* 5: 360–364.

Danovaro, R., S. Fonda Umani, and A. Pusceddu. 2009. "Climate Change and the Potential Spreading of Marine Mucilage and Microbial Pathogens in the Mediterranean Sea." *PLoS ONE* 4: E7006. doi:10.1371/Journal .Pone.0007006.

Dello, K. 2007. "Trends in Climate in Northern New York and Western Vermont." Master's thesis, State University of New York at Albany.

Frumhoff, P. C., Mccarthy, J. J., Melillo, J. M., Moser, S. C., and Wuebbles, D. J. 2007. "Confronting Climate Change in the US Northeast: Science, Impacts, and Solutions. Synthesis Report of the Northeast Climate Impacts Assessment (NECIA)." http://www.northeastclimateimpacts.org/.

Harlow, W. M., E. S. Harrar, J. W. Hardin, and F. M. White. 1996. *Textbook of Dendrology.* New York: McGraw-Hill.

Hayhoe, K., et al. 2007. "Past and Future Changes in Climate and Hydrological Indicators in the U.S. Northeast." *Climate Dynamics* 28: 381–407.

Hulme, M., et al. 1999. "Relative Impacts of Human-Induced Climate Change and Natural Climate Variability." *Nature* 397: 688–691.

IPCC. 2007. "Summary for Policymakers." In: *Climate Change 2007: Impacts, Adaptation and Vulnerability. Contribution of Working Group II to the Fourth Assessment Report of the Intergovernmental Panel on Climate Change,* M. L. Parry et al., eds. Cambridge, UK: Cambridge University Press.

Keim, B. D., et al. 2003. "Are There Spurious Temperature Trends in the United States Climate Division Database?" *Geophysical Research Letters* 30: 1404, doi:10.1029/2002GL016295.

Lovett, G. M., and M. J. Mitchell. 2004. "Sugar Maple and Nitrogen Cycling in the Forests of Eastern North America." *Frontiers in Ecology and the Environment* 2: 81–88.

McKibben, B. 1990. *The End of Nature.* New York: Penguin Books.

———. 2002. "Future Shock: The Coming Adirondack Climate." *Adirondack Life,* March–April.

———. 2006. "A Win-Wind Situation." *Adirondack Life,* September–October.

———. 2009. "Half-Precessional Dynamics of Monsoon Rainfall Near the East African Equator." *Nature* 462: 637–641.

Muller, E. H., L. Sirkin, and J. L. Craft. 1993. "Stratigraphic Evidence of a Pre-Wisconsinan Interglaciation in the Adirondack Mountains, New York." *Quaternary Research* 40: 163–168.

New England Regional Assessment. 2001. "Preparing for a Changing Climate: The Potential Consequences of Climate Variability and Change; A Final Report." Durham: University of New Hampshire. http://www.necci.sr.unh.edu.

Prasad, A. M., and L. R. Iverson. 1999. "A Climate Change Atlas for 80 Forest Tree Species of the Eastern United States (Database)." Northeastern Research Station, USDA Forest Service, Delaware, Ohio. http://www.fs.fed.us/ne/delaware/atlas/index.html.

Schiermeier, Q. 2010. "The Real Holes in Climate Science." *Nature* 463: 284–287.

Meehl, G. A., et al. 2007. "Global Climate Projections." In: *Climate Change 2007: The Physical Science Basis. Contribution of Working Group I to the Fourth Assessment Report of the Intergovernmental Panel on Climate Change*, S. Solomon et al., eds. Cambridge, UK: Cambridge University Press.

Spaulding, P., and A. W. Bratton. 1946. "Decay Following Glaze Storm Damage in Woodlands of Central New York." *Journal of Forestry* 44: 515–519.

Stager, J. C., and M. R. Martin. 2002. "Global Climate Change and the Adirondacks." *Adirondack Journal of Environmental Studies* 9: 1–10.

—— et al. 2009. "Historical Patterns and Effects of Changes in Adirondack Climates Since the Early 20th Century." *Adirondack Journal of Environmental Studies* 15: 14–24.

Stine, A. R., P. Huybers, and I. Y. Fung. 2009. "Changes in the Phase of the Annual Cycles of Surface Temperature." *Nature* 457: 435–440.

Thaler, J. S. 2006. *Adirondack Weather: History and Climate Guide.* Yorktown Heights, NY: Hudson Valley Climate Service.

Thomson, D. J. 2009. "Shifts in Season." *Nature* 457: 391–392.

Trombulak, S., and R. Wolfson. 2004. "Twentieth Century Climate Change in New England and New York." *Geophysical Research Letters* 31: L19202, doi:10.1029/2004GL020574.

Willis, K. J., and S. A. Bhagwat. 2009. "Biodiversity and Climate Change." *Science* 326: 806–807.

Index

Italic page numbers refer to photographs and charts.

carbonate molecules, 105
carbonate rocks, reactions with carbon
dioxide, 36–38, 103
carbonic acid
in the ocean, 103
in rainwater, 37
Cazenave, Anny, 122
cells, body, replacement of, 100
Cenozoic era, 68–69, 78–83
central Asia, climate changes expected
for, 223
chalk, 36
change, fear of, 239
Chiarenzelli, Jeff, 219
China
ancient, 60
climate changes expected for, 207
limestone features in, 38
chlorophyll, 5
Chukchi Sea, 158
Churchill, Canada, 157
Churchill, Winston, quoted, 67
clathrate-gun hypothesis, 74
clathrates, 74, 75
clays, 76
climate change
and absence of refuges to escape to,
146
animals migrating in response to,
61–62, 66, 82–83, 146–47, 150–55, 225
calls to "take action," 200
commitment to, 34
cooling direction of, in recent
epochs, 29
"don't panic and don't give up,"
228–31
doubters and deniers of, 125, 148,
216–18
fatalistic do-nothing approach to, 48
flexibility in adapting to, 201–2
and geological changes to the land,
154
greenhouse gases and, 8–9, 65–66, 71
humans adapting to, 121, 157, 185–86
nonhuman causes of, 21–26

politicized discussion about, 143
possible benefits of, 102, 156–60, 197
uncertainty about, 130
whiplash effects in, 30–31
winners and losers from, 202, 239–40
See also Anthropocene epoch; global
cooling; global warming
Climate Change Institute, University of
Maine, Orono, 22
climate control
consciousness-raising about, 233
controlled burning of coal, suggested
for, 234–35
obstacles to, from politics and media,
233
climate cycles
computation of, into the future,
23–26
overlapping and interacting, 22–26
climate models, computer-simulated,
31–42
checked against geohistorical studies,
49
El Niño, 199
Greenland, 123, 166, 171
Northeast U.S., 221–22
regional assessments, 209
tropical mountains, 196
climate prediction, for the deep future,
9–12, 23–26
climate system, variability and
instability of, 23
Climate Wizard, 223
CLIMBER climate model, 32, 49
Clovis-age cultures, 66
coal
burning of, as suggested climate
control measure, 234–35
CO_2 released from burning of, 6–7
formation of, 6
coasts
cities on, sinking of, 137–38
collapse of, 156
population living on, effect of sea
level change, 125, 133–35

native settlements, Arctic, threatened
by sea ice melt and coastal collapse,
156
natural world
end of, as distinct from humanity,
1, 8
humans' position in, 8
naturalists, amateur, observations of,
218–20
Nature magazine, quoted, 203
Neanderthals, 63–64
New England Regional Assessment
(NERA), 210
New Orleans, 137
New York City, climate changes
expected for, 224
New York Harbor, 224–25
New York state
upstate, climate changes expected
for, 208–22
weather records, 216
New Zealand, ancient, 62
Nile River, 188, 197
nitrogen, 88, 95, 237
nitrogen fertilizer, extent of use, 4
Noah's Flood, 118, 126
Norsemen, 162–63, 172
North America
ancient, 60, 62, 239
climate changes expected for, 223
North Atlantic
deep coral reefs, 109–10
lava rifts in, during Cenozoic, 73
North Atlantic Oscillation, 25
north polar cap
ice-free future, 139, 142
life on and under, 143–53
recovery of, in future global cooling,
160–61
North Sea, 133
"Northeast Climate Impacts
Assessment" (NECIA), 215, 222
Northeast U.S., climate models, 221–22
northern Europe, climate changes
expected for, 223

Northern Hemisphere, ice ages in,
22–23
Northwest Passage, modern opening of,
142, 158
Northwest Territories, 146–47, 159
Norway, 163, 181
territorial claims, 158, 164
Nova Zemlya, 176
nuclear testing in the atmosphere, 98
Ny Fjord (of future Greenland),
176–79
Nyanza Project, 90

observers, naturalist, 218–20
ocean currents, 17–21
ocean sequestration of carbon, 117
oceans
acidification of, by CO_2, 10, 34, 40,
72–73, 77, 102–17, 152
deep, temperature of, 76–77
food webs, 107–8, 114–16
recovery process, length of time
required, 106
uptake of carbon dioxide by, 36,
103, 115
upwelling zones, 115–16
See also coasts; sea level
Oerlemans, Johannes, 141
Oligocene epoch, 3
Opdyke, Neil, 151
orcas, 150
O'Reilly, Catherine, 91–92
Overpeck, John, 127
oxygen, 237–38
primordial release into the
atmosphere, 5–6
oxygen-18 dating, 69
oysters, 58, 77
ozone, 6

Pagani, Mark, 70–71
Pakistan, 197
Paleocene-Eocene thermal maximum
(PETM)
abrupt onset of, 83–84

Russia
  Arctic Ocean fishing, 152
  fossil fuel resources, 160
  territorial claims, 158–59
Rutgers University, 237

Saga Rauda (Eirik's saga), 162
Sahara, 190
  ancient, 188
Saint Lawrence River, 178
Saint Regis Mountain, N.Y., 14–15
salt marshes, 135–36
Sandweiss, Dan, 191–93
Sangamon (Eemian) interglacial, 54
Santa Claus, home of, 176
satellite systems, geographic
  measurement by, 120–21
"Save the Carbon" slogan, 235
Schelske, Claire, 89–90
Schiermeier, Quirin, 157, 223
Schmittner, Andreas, 41
Schmitz, Matt, 195
scientists
  climate activists among, aggressive
    stance of, 240–41
  popular distrust of, 230, 241
  skeptical mindset of, 130, 216
sea ice, thickening of, 139
sea level
  23 foot rise (moderate scenario),
    40–42
  230 foot rise (extreme scenario), 34–35,
    123
  Britain's brass sea-level benchmark,
    119–20
  defining and measuring, 118–21
  duration of rise, 134–35
  fall of, from global cooling, 121–22,
    135–36
  during interglacials, 56–59
  during last ice age (400 feet below
    present), 121–22
  maps of change, 127–29, 133
  oceanfront settlements affected by,
    125, 133–35, 137–38

  of past geological eras, 56–59, 65, 84
  rise in 20th century (7 inches), 122,
    125
  rise in 21st century, projected (1 to 2
    feet), 122, 125, 130
  rise of, from global warming, 11, 65,
    84, 118–21, 122, 125, 130, 166
  speed of rise, 122–26, 129–30, 132
seasonal solar heating, 51
seawater
  chemistry of, 34, 105
  cold, sunken floods of, from
    Antarctica, 108–9
sediment cores
  marine, 73
  from PETM era, 69–70
Shanghai, 138
Shea, Judith, 217
shells, marine, 105
  attacked by CO_2, 36
  building and maintaining of, 38,
    104–6, 111
silicate rocks, 39
Sirocko, Frank, quoted, 13
Smol, John, 154
snow
  predicting, 222
  survival of, in a warming world, 76,
    223–24
  and tourism, 210
South Pole, 139
southern Africa, 198
  ancient, 64
  climate changes expected for, 204–7
  weather system of, 206–7
southern Australia, 207
  climate changes expected for, 223,
    226
Southern Hemisphere, ice ages in, 23
Southwest, American, climate changes
  expected for, 223, 226
species, invasive, 226, 231–32
species extinction
  in Anthropocene epoch, 4, 12, 62–63
  future, predicted, 10

from global warming, 12, 77
irreversibility of, 102, 113–15, 117
local vs. global, 225
and loss of biodiversity, 153–54
from loss of climate zones, 226
Stager, Curt (author)
African lake studies, 188
in high school physics class, 119
Lake Nyos observer, 45–47
professional background of, 2
radio show, 219
upstate New York home, 208, 218
starfish, 114–15
Steffen, Konrad, 165
Stern, Sir Nick, 184
Stickley, Catherine, 160
Stine, Alexander, 208
Stoermer, Eugene, 4
stomatal index data, and CO_2 concentration, 54
Stone Age, 9
Stott, Lowell, 72
Sudan, 197
Suess, Hans, 88
Suess effect, 88–93, 97–99
sugar maples, 220–21
sun
eleven-year cycles of, 197–98
energy of, 5–6
sunlight (insolation)
high-altitude deflection of, proposed, 116–17
surface of Earth exposed to, 21–25, 24
super-greenhouse scenario, 67–85
Svalbard archipelago, 142, 145, 148, 176
Swiss Federal Institute of Technology, 237

Tanzania, 197
ancient, 71
Tchernov, Dan, 112
temperate zone
climate changes expected for, 203–27
warm/cold cycles in, 186

temperature, global mean
3 to 7 degree Fahrenheit (moderate) rise of, 34–35
9 to 16 degree Fahrenheit (high) rise of, 40–42
ancient, estimating, 55
drop in, from high levels, 38–39
local divergences from pattern of, 212–15
rise of, in Cenozoic, 68–71, 70
temperature, sea-surface, rise in, 69–71, 76
tetraether lipids, 70–71
Texas, 199
Thaler, Jerome, 214, 217
thermohaline circulation (THC), 17–18
thermonuclear war, and extinction threat, 44
350.org, 33, 233
Thule (Qaanaaq), 174
tigers, 136
tilt cycle, 22
time, deep, beyond 2100 AD, 1–3
Tingle, Alex, 128
Tokyo, 137
trawlers, fishing, 111
Treaty of Kiel (1814), 164
trees
ancient, 78–80
northward migration of southern types, 60, 172, 211, 224
optimal temperature ranges of, 220–21
tropics, the, 181–202
temperature rise in, 184–86
weather system of, 188–90, 196–99
wet/dry cycles in, 186–88
Tropics of Cancer and Capricorn, 189
trout, 224
Tuktoyaktuk, Alaska, 157
tumors, 100
tundra, 153, 156
Turkana people, 181–82